Teacher's Guide

Shelterwood

DISCOVERING the FOREST

Judy Kellogg Markowsky

Illustrations by
Rosemary Giebfried

TILBURY HOUSE, PUBLISHERS
Gardiner, Maine 04345

TILBURY HOUSE, PUBLISHERS
2 Mechanic Street #3
Gardiner, Maine 04345

Design & Layout Rosemary Giebfried, Bangor, ME
Illustration Rosemary Giebfried, Bangor, ME
Editorial & Production Jennifer Elliott
Barbara Diamond
Printing & Binding InterCity Press, Rockland, MA

□

First edition, September 1999.

10 9 8 7 6 5 4 3 2 1

Table of Contents

Classroom Activities and Student Handouts

Student Handout Pages: *Fun fact sheets, quizzes, and educational puzzles about forests*

Student Handout Pages: *Fun fact sheets, quizzes, and educational puzzles about forests*

Student Handout Pages: *Fun fact sheets, quizzes, and educational puzzles about forests*

Many Thanks! Thanks to friends and colleagues in the Maine Audubon Society: staff and volunteers, forest advocates, community leaders, environmental educators, and forest policy analysts who have worked with Maine legislators, forest owners, and stakeholders to build a consensus and change forest practices gradually in a climate of mutual respect. Thanks also to those Maine Audubon staff and volunteers who have taught our younger generation about the natural, economic, and social values of the forest and other natural environments in Maine.

Thanks to friends and colleagues at the University of Maine's College of Natural Sciences, Forestry, and Agriculture, especially its Department of Forest Management and its Department of Wildlife Ecology.

Thanks to Alan Kimball of the Department of Forest Management, College of Natural Sciences, Forestry, and Agriculture, University of Maine, Ted Shina of the Old Town Timber Group, Fort James Corporation, and Joel Swanton of Champion International Corporation for reading and making helpful comments on the manuscript. Also thanks to Mel Ames and Max McCormack for their helpful comments. These are all Maine forestry professionals with a strong commitment to education.

Thanks to faculty and students in the University of Maine's College of Education, particularly the Science and Environmental Education program, who have helped develop, test, or evaluate many of the ideas and activities in this book.

Thanks to colleagues and friends in the Maine Department of Inland Fisheries and Wildlife, and also to local staff of the U. S. Fish and Wildlife Service.

Thanks to Rosemary Giebfried for splendid illustrations and design, for unflagging enthusiasm about the project despite multiple challenges, and for her knowledge of computers.

Thanks to Jennifer Elliott and Susan Shetterly for their collaboration on this project.

Thanks to family for their support and encouragement.

Judy Kellogg Markowsky

Barred owl feather

CHAPTER ONE

What Is a Forest?

Ask your students, "What is a forest?" You might get a response like, "A place with trees." You can ask further, "The schoolyard has trees. Is that a forest?" Usually that doesn't match the mental image. "No…it has to be a big area, with a lot of trees, so you can't see through it. It has to feel wild."

Most would agree, a forest has to be a big, undeveloped area with lots of trees. It should look and seem wild, unsettled, and natural.

But a forest has much more than trees! It has a special balance of air, soil, water, sun and shade, wildlife, life and death that is fascinating to study.

Going into a forest—and learning about it—is an adventure. It is often dark inside the forest, and, from a child's point of view, there are many hiding places for animals that you can't always see. Perhaps that is why some people are fearful of forests, preferring the bright sunlight of open fields, lawns, and gardens. History and progress often seem to consist of clearing away the forests for agriculture or for building cities. For many, our forests now symbolize the natural landscape that we are in danger of losing.

Your students will love learning about the forest, with its mysteries and adventures. Through *Shelterwood*, they will learn about forest stewardship, in which a woodlot is maintained largely in its natural state even as it is utilized for products. They will also learn some important concepts in wildlife biology.

Children can learn about the forest through science and literature, through classroom exercises, through reading and math, through their parents, through the World Wide Web, and, perhaps best of all, through their own experience.

Children love experiencing the forest in a hands-on, physical way. They love climbing over the trunk of a fallen tree, peering into a hollow log, feeling the mosses and lichens on a tree trunk, feeling the bark texture, climbing through a blowdown, crawling under a thicket. A carefully prepared trail, and adult supervision, can safely provide children with these experiences.

> *Wherever we go in the wilderness,*
> *we find more than we seek.*
> *-John Muir*

Children love tapping on a tree (as a woodpecker does) and listening carefully to determine whether the wood is hollow, rotten, or sound and solid. They love measuring a tree by wrapping their arms around it. It takes several children to wrap arms around a large tree! Later, back in the classroom, they can measure their armspans and approximate the tree's circumference, then its diameter. Math is more fun when it's related to an experience.

Children love studying a cone closely, feeling the pattern of the scales and seeing the imprint of the seeds on the underside of each scale. They love feeling and even sniffing the scent of the needles or leaves. In the Northeast, fir needles have a wonderful smell, while white spruce stinks! In the West, the bark of a Ponderosa Pine smells like vanilla.

Such concrete experiences can lead the children into reading about the forest, checking Web sites, writing and doing math about the forest, and other abstract learning. This Teacher's Guide shows you how.

Name

Woodland Wanderers

Life in the forest takes wondrous and amazing journeys. Can you guess what makes each journey?

•Draw a line that connects the little traveler to its journey.

Name

Fruits of the Forest

Almost all parts of plants are eaten by one animal species or another. Fruits, seeds, leaves, twigs, bark, stems and roots all provide food to different kinds of animals.

- **Use the words from the Word Box to answer the Crossword Clues.**
- **Cross off each word as you use it.**

Across

3. Long, dangling flowers found on birch and other trees.
5. Small, juicy fruits.
6. Beavers, porcupines, and rabbits eat this tough outer part of trees.
7. Apples and oranges are kinds of ------
9. Pine trees store their seeds in these woody structures.
10. These fall from oak trees, chestnut trees, beech trees and hickory trees.

Down

1. Squirrels like to bury these in the ground.
2. Another kind of nut that animals love.
4. The winged seeds of maples, ash, and elm trees.
8. The underground part of a tree that absorbs water and minerals.

a birch leaf and catkin

Word Box

acorns	catkins	roots
nuts	cones	fruits
samara	hazelnuts	berries
	bark	

maple samara

Oh, nuts!

Nuts are one of the most important foods available to wildlife because they are very nutritious. Some nuts that are common in the forest are acorns, pecans, and beechnuts. Only squirrels and chipmunks are able to eat hickory nuts and butternuts because the shell is too thick for the other animals to break open.

Did you know?

Common weeds like crabgrass, ragweed, and pigweed are extremely valuable food sources for animals because they have so many seeds. For instance, pigweeds have 100,000 seeds per plant!

Name

Forest Friends

There are a lot of animals in the forest.
Most of our forest friends are secretive and nocturnal, and are not always easy to see during the day.

- Use the words from the word box to see how many you can find!

Word Box

bat	hawk	flying squirrel
bear	lynx	salamander
beaver	moose	shrew
bluejay	mouse	snake
caribou	owl	snowshoe hare
chipmunk	porcupine	toad
coyote	raccoon	turtle
deer	red squirrel	wolf
fox		woodchuck

L P K N E N I P U C R O P B Z
E T C Z G K W A H R E V A E B
R Y U N W N A L P E D T M T Z
R G H S T U B N Q D S M E O W
I I C H O M J Y S N Q O W A D
U U D R J P B A C A U U O D Y
Q N O E K I X J C M I S E M T
S N O W S H O E H A R E L J N
G C W H A C F U B L R A M N Q
N O Z D X N Y L O A E I E O S
I P J A E G D B C S L Y B B I
Y D E T O Y O C C V T V T O Q
L R A U M W O L F C R G I Q U
F L U M L O A I H G U Z D V I
S D O Z N A C Z C H T E B Y A

Chapter Two
What Kind of Forest?

Hardwood or Softwood?

In the U. S. there are two major categories of trees, often called "softwood" and "hardwood." Softwoods are also commonly called evergreens, needle-leaved trees, or conifers. Hardwoods are also commonly called deciduous trees, or broad-leaved trees. A hardwood forest consists mostly of hardwood trees, and a softwood forest consists mostly of softwood trees.

The wood of most hardwood trees *is* harder than that of most softwoods.

The seeds of most hardwoods, instead of forming inside a cone (a simple structure), are formed inside a flower (a more complex structure).

In the context of science, softwoods are called "gymnosperms" (which means "naked seed") and hardwoods are called "angiosperms" (which means "seed in a vessel").

Examples of softwoods are: pine, spruce, fir, hemlock. Examples of hardwoods are: oak, maple, beech, hickory, sycamore. There are many species for some of these types.

Many forests have both hardwood and softwood trees. If neither dominates, it is called a "mixed forest."

The maps on the Student Handout (page 15) can help you figure out what forest type surrounds you. Your eyes and ears can help, too!

The kinds of forest look very different throughout the seasons. Do you live where the leaves change color and then fall to the ground? The Northeast is known for its vivid fall colors. But in the South, some broadleaf trees keep their leaves all winter.

Most softwoods keep their needles all winter. But needles fall, too! Look carefully. Every fall, older pine needles drop off, leaving the newer needles, at branch tip, on the tree all winter. You can always find the older, fallen needles at the base of the tree.

The forest is far more than a collection of trees.
–Gifford Pinchot

Most softwoods are cone-shaped when seen from a distance. Longer branches at the bottom, shortening upwards toward the top of the tree, give it this cone shape. This may be an adaptation that helps shed snow in snowy climates. Snow lands at the end of each branch, dispersing the snow all up and down the tree, not putting too much weight on any one branch.

There is one kind of coniferous (softwood) tree that drops its needles in the fall, leaving a bare tree-skeleton behind all winter. In spring, beautiful chartreuse needles emerge. You read about this tree in the book *Shelterwood*. It's called a hackmatack. (Some people call it a larch or tamarack.)

Activity 1: What Forest Do YOU Live in?

Objective: Children will learn the major forest types, especially the one of their region.

Concept: There are different forest types in different broad regions of North America: the Northeast, the Southeast, the Midwest, the Rocky Mountains, and the Pacific Northwest.

You Will Need: •*The simplified map of the major forest regions of the U. S. on page 15*

What to Do: Show students a map of the forest regions, have them find where they live, and see which forest type they live within. *Where can they go to see an example of that forest type?* If your school and homes are in a large developed area, the trees nearby may well be non-native plantings which do not represent the original forest type of the area.

Macbeth: Who can impress the forest, bid the tree / unfix his earth-bound root?
I will not be afraid of death and bane / till Birnam forest come to Dunsinane.
Messenger: As I did stand my watch upon the hill, and looked toward Birnam,
The wood began to move.

–William Shakespeare

Activity 2: Why Do Some Regions Have NO Forest?

Objective: Children will learn the importance of rainfall in determining whether a desert, grassland, or forest would occur naturally in their area.

Concept: Ecologically, it's usually a matter of water. It takes a lot of water to grow a tree! Instead of forests, there are large deserts in areas with very little rainfall. Grasslands occur naturally over large areas with more rainfall than deserts, but less rainfall than forests. (In the West, mountains can be "cloud-catchers," so they are sometimes forested even though surrounded by desert.)

You Will Need: •*The simplified rainfall map on page 16*

What to Do: Show students the simplified rainfall map. *How much rainfall occurs in YOUR area per year?* That's what would determine whether your area would naturally have desert, grasslands, or forest. (Of course, in developed areas, people have often changed these generalizations by irrigation, or by clearing the forest, then planting for agriculture or lawns.)

Name

The Great American Forests

The Northern Forest

The Southern Forest

The Central Hardwood Forest

The Rocky Mountain Forest Region

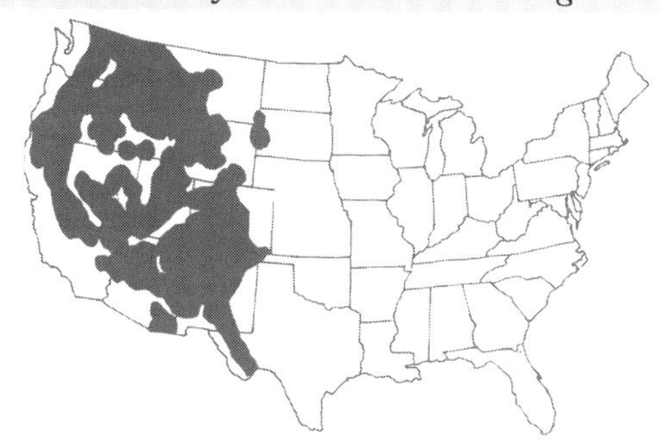

The Pacific Coast Forest

Sensing the Forest

Shh-h-h-h! What quiet tree noises have you heard in your forest? The rustle of leaves, the creaking of one tree leaning on another? The sighing of wind through pine needles?

Is the forest floor soft to walk on? Or dry and crunchy? Or wet and squishy?

Make an X on the map where you live.

Combined Forest Regions

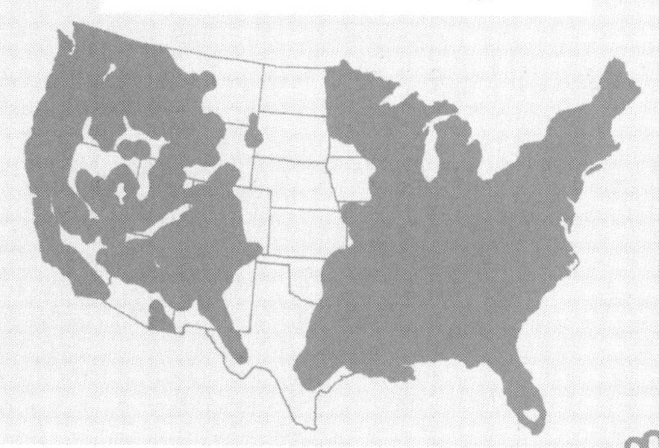

What forest type do you live in?

1. The **Northern** Forest
2. The **Southern** Forest
3. The **Rocky Mountain** Forest Region
4. The **Central** Hardwood Forest
5. The **Pacific Coast** Forest

Name

Reviewing the Rainfall

Instead of forests, there are large deserts in areas with very little rainfall. Grasslands occur naturally over large areas with more rainfall than deserts, but less rainfall than forests.

- **Look at the simplified Rainfall Map below.**
- **Make an "X" on the map in the area where you live. Then answer the questions below.**

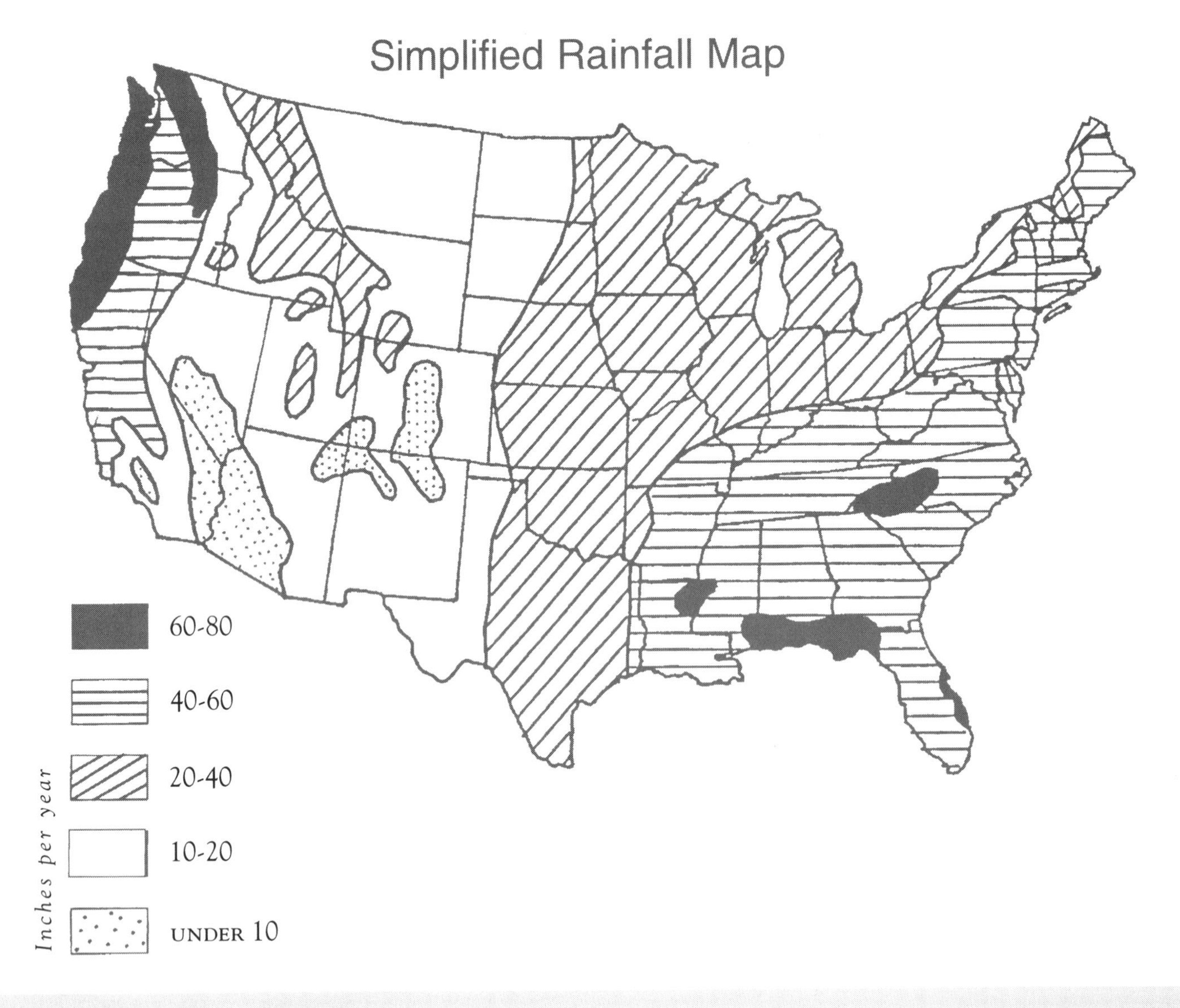

What's it like where YOU live?

1. How much rainfall occurs in your area each year?_______ inches per year.

2. Is the area where you live forested, or is it grassland or desert? _______________

3. Does the amount of rainfall your area receives correspond with the region (biome) you live in?_______

Name

Tree Trivia

Trees will always be one of the most important natural resources we have. You can help in their conservation by planting trees in your garden or at school, and by recycling old newspaper.

•Each sentence has a different extra letter that doesn't belong.
•Cross out the extra letter wherever it appears in the sentence.

1. Kone treek kcan providke a dayk's koxygen fork up to fourk peopkle.

2. Youngj bajrk is smoojth. Oldj bajrk crajcks and flajkes.

3. Countingp the prings on a sepction of ptrunk can tellp us the agep of a treep.

4. Thxere arxe axbout 1000 xspecies of trxees in Norxth Americax.

5. Treesc are the largcest licving thicngs in the wocrld.

6. A qyoung treeq is qcalled a qsapling.

7. Barkv grovws frovm the invside and pusvhes the volder barvk outwvard.

8. Levery linch arolund the girlth of a treel corrlesponds to a yearl in a treel's growlth.

9. Dmore thand 400,000 leadves cdan fadlls from a sindgle dlarge dtree.

10. Theb tallbest treeb in the bworld is a Califbornia redbwood and is obver 360 feebt btall.

11. Thef bristleconfe fpine is the foldest lifving treef at ofver 4,600 yearfs fold.

We can't live without trees.
Trees and plants convert carbon dioxide and water to carbohydrates and oxygen.
Without oxygen, we couldn't live.

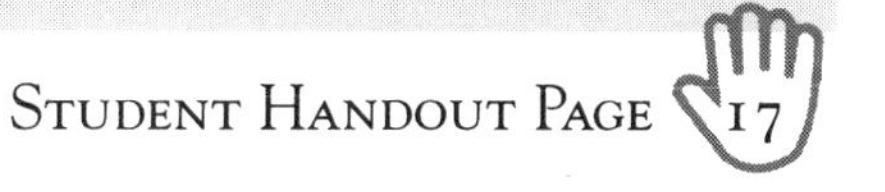

Name

Crack the Code

Learning the names of all the different kinds of forest trees can be tricky. Here's a joke that's just for fun.

- **Try to find out the answer to this Forest Riddle.**
- **Use the code to fill in the letters. The first letter has been done for you.**

Riddle: What kind of pine has the sharpest needles?

Answer:

A

Clue Box

A =

I = 

O = 

U =

E =

N = 

C =

P =

R = 

Conifers are trees that have cones.
Pines, hemlocks, spruces, and firs are familiar examples. Most conifers also are evergreen and have needles. Compare their needles. On some pines, the needles are SHARP like the quills of a porcupine–ouch! And some pine trees have needles that are soft and pliable and pleasant to touch, feel, and even smell.

Activity 3: Learn to See Form and Patterns in the Trees

Red Maple

The trunk of a hardwood tree goes up a ways, then has branches radiating upwards.

White Pine

The central trunk of a softwood tree has branches coming off it regularly, getting smaller toward the top.

Objective: Children will learn to interpret a tree's structure.

Concept: Hardwood and softwood trees often have a different structure when you learn to look for it. Many hardwoods have a trunk that goes up a ways, and then has branches radiating upwards. Many softwoods have a structure of a central trunk with branches coming down off it regularly, getting smaller toward the top.
Can you explain why this is?

You Will Need: •*Clipboards and art materials that can be used outdoors*

What to Do: Have the children try to ignore the leaves or needles and only sketch the branches of the tree. Their sketches will represent the tree's structure.
It's best if they interpret the tree as they see it. You would not want to be dogmatic about what a softwood or hardwood is "supposed" to look like, because there are exceptions to the generalization sketched above.
But the structure of a tree often is a clue to its identification, and sometimes, to the conditions under which it grew.

What kind of tree? Look at its shape.
Hardwood trees are usually rounded. Softwoods often have a triangular shape.

Activity 4: Learning to See Patterns in Twigs and Branches

Objective: Children will learn the difference between "alternate" and "opposite" leaf and branch patterns.

Concept: "Alternate" and "opposite" leaf and branch patterns are easy to learn and observe, and are an important clue to tree identification.

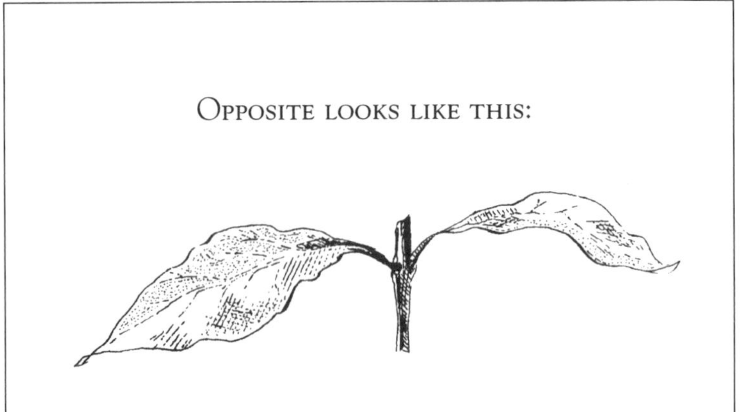

You Will Need:

- *Clippers to cut about 30 six-inch twigs*
- *Two containers in which to put the twigs*

(Be sure none of the twigs are poison ivy, oak, or sumac.)

What to Do: Explain and demonstrate a branch with alternate leaves or buds, and another with opposite leaves or buds. Then let the children sort the twigs into the two containers.

Note: The opposite pattern is uncommon in trees, although ash and maple have it. It is, however, a very common pattern in shrubs, which are harder to learn and identify than trees.

Activity 5: Learning to See the Difference Between Trees and Shrubs

Objective: Children will learn to differentiate trees from shrubs by learning to observe their typical woody structure.

Concept: Shrubs and trees both have woody stems. A shrub typically is shorter and has many thin stems. A full-grown tree typically will be taller and have one main trunk.

You Will Need: •*Construction paper and scissors*

What to Do: Explain to children the difference between trees and shrubs, and show them they can make this pattern with construction paper. Then, outside, see if they can find good examples of each.

Note: They might occasionally find some "in-between" forms that are challenging to interpret. Such is life.

He who plants a tree
plants a hope.
–Lucy Larcom

Activity 6: Keep a Tree's Diary

Objective: Students will get practice in tree identification, patient observation, creative writing and art, and keeping a journal over time.

Concept: Trees change all the time, but seeing these changes requires close observation over a period of time.

You Will Need:
•*Clipboards or notebooks*
•*The Student Handout Sheet on page 22 may also be used to help gather data*

What to Do: Have each student identify a tree in his/her backyard or schoolyard. They will have weekly observation periods, and write in their journals the date, time, temperature, weather, and their observations. Different angles and themes can be brought in throughout the year, such as writing from the tree's point of view: "Because it rained all morning, droplets dripped down my bark and off my twigs at 1:00 PM." The possibilities are endless, limited only by your creativity and that of the children. Examples: write a poem; measure the growing leaves weekly in spring; look for insects hiding in the bark; dig among the roots, looking for the tree's soil (see Chapter 6 for soil information and activities). At the end of the year, ask the students how the tree has changed.

Name

Tree Diary Worksheet

Identify a tree in your backyard or in the schoolyard.

- **Observe your tree each week. Spend a few minutes watching your tree carefully.**
- **Fill in the date, time, temperature, weather, and your observations.**
- **Use this sheet three to four times this year to get seasonal differences.**

Weekly Observation Chart

Date	Time	Temperature	Weather Conditions	What do you see? What do you hear?

What's the width of your tree? Measure the circumference of the tree's trunk with a tape measure. Write your answer here_________________.

Write about what you smell... flowers, bark, leaves, fruit.

Trace one of the leaves.

Describe the location of your tree:

____

____

Make a rubbing of the bark. Holding a piece of paper against the bark, scribble over the paper with the side of a crayon. Is the bark thick and rugged? Or thin and smooth?

Sketch any flowers and fruits from your tree.

What is the area around your tree like?

Write a short poem about your tree.

Draw a picture of your tree.

CHAPTER THREE

Learning to Identify Trees

Too often, people try to learn or teach too many tree species at once. Then, learners get confused, give up, or forget everything. Much better is to learn about just a few tree species at a time, over an entire year of learning.

First, learn three distinctive trees in your neighborhood and visit each kind often, repeating its name, noting its important characteristics, and observing how it changes over time. Then, try to learn two more, and keep visiting all of them until you know them really well.

In neighborhoods, towns, parks, and cities, many of the trees will be non-native, planted species. They may be different species from those in a nearby forest, adding to the confusion when you're starting out.

To get started, perhaps a parent or friend who is a step ahead of you in knowledge of trees can help. Sometimes a town or neighborhood beautification committee member, landscaper, or forester is knowledgeable about trees and can help. The important thing is to get really familiar with several kinds right in your own schoolyard or neighborhood.

> *"The wonder is that we can see these trees and not wonder more."*
>
> *-Ralph Waldo Emerson*

It's probably best to learn trees, and their fascinating seasonal changes, all year, not just in a two-week "unit." When first learning trees, most people learn them by their leaves, the easiest way to start.

But the more you learn about trees, the easier it becomes to learn more! You'll find you can identify them by their bark, by their shape and branching structure, by their leafless twigs; sometimes even by their smell, or by the sound the wind makes blowing through them. Does the wind blowing through needles sound like a sigh, or a whistle? Do the leaves rattle together, or just click lightly, when the wind blows through?

Then you can reinforce "tree learning" all year. That's important because too often, trees are taught too much at once, and are forgotten afterwards.

Sycamore trees were considered sacred in ancient Egypt and are the oldest trees represented in art.

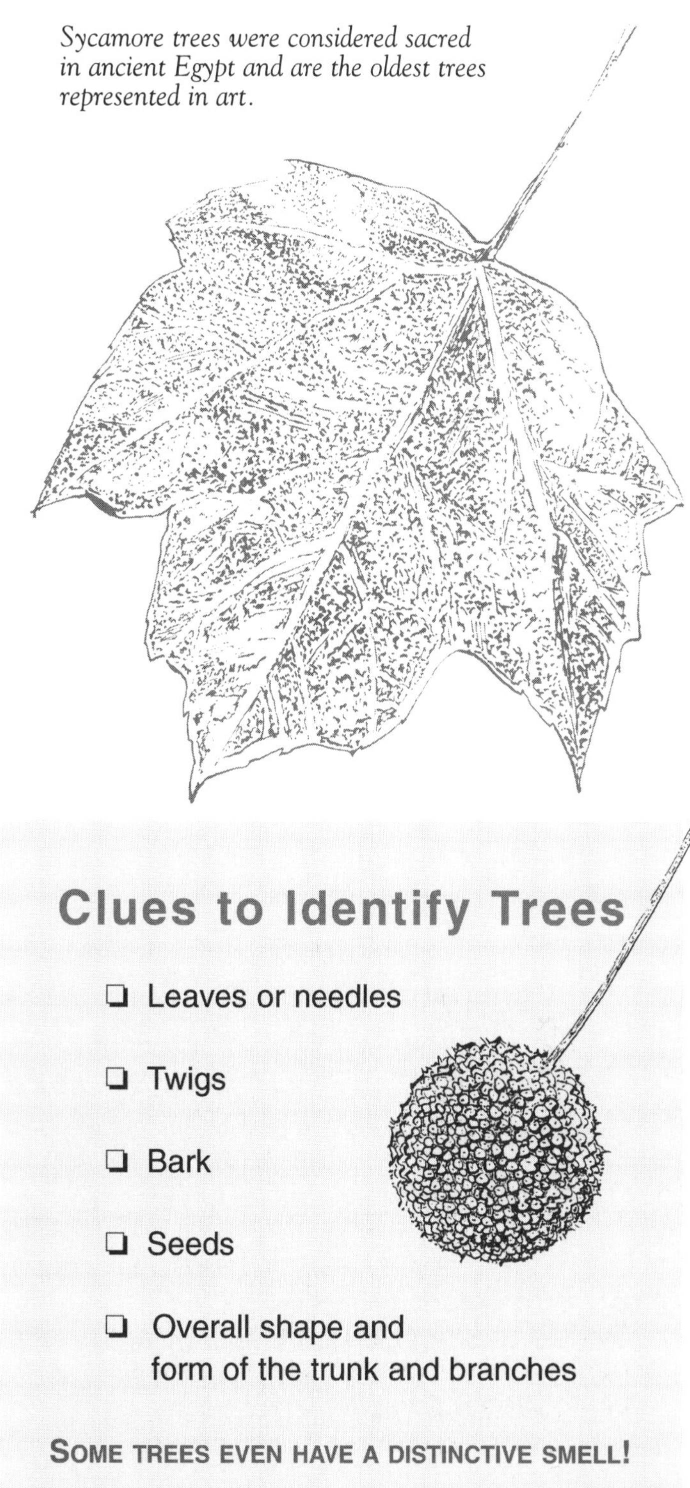

These activities feature several distinctive kinds of trees found across the country. Not all trees are distinctive; birch and cherry trees, for example, have simple, nondistinctive leaves that can be hard to tell apart when you are first learning trees.

Activity 1: Find a Pine

(For teachers and students who are first learning to identify trees.)

Objective: To learn to identify a pine.

Concept: Pines are found throughout the U. S. and Canada. There are many different species of pine, but chances are several live near you. All pines have long, fine needles that grow in bundles.

You Will Need:
- *Clippers*
- *If necessary, permission to clip off a branch about 2 feet long*
- *A heavy, sturdy, non-tippy vase with water*
- *The Pine Tree Identification Chart on page 25*

What to Do: Cut a 2-foot pine branch, with cones if possible, and put it in a vase indoors. Study its features over and over again. There are many species of pine; you may need local assistance to learn what species are common in your area and how to differentiate among them. When you learn the name of your pine, children can make a label for it. Children can make their own **"Field Guide to Local Pines,"** showing in writing and pictures the identifying features of each species.

Activity 2: Find a Maple

Objective: To learn distinguishing features of a maple tree.

Concept: Maple trees are also found throughout the U. S. and Canada. There are numerous kinds, but nearly all have "palmate" leaves—the veins radiate out from the leaf's "wrist" or stem. The shape of a maple leaf is on the Canadian flag.

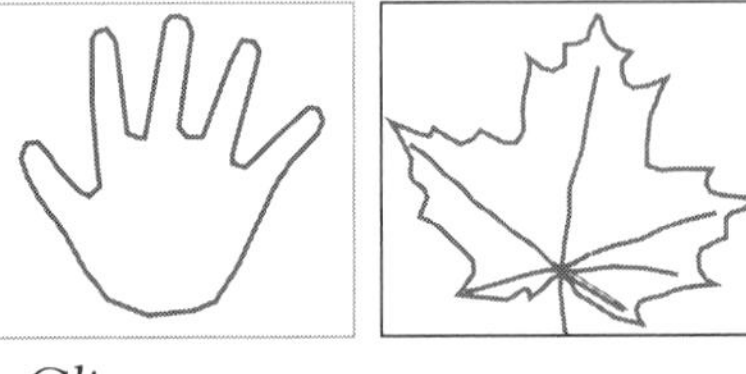

Most maple leaves are "palmate"—the veins radiate out from the leaf's "wrist" or stem.

You Will Need:
- *Clippers*
- *If necessary, permission to clip off a branch about 2 feet long*
- *A heavy, sturdy, non-tippy vase with water*
- *The Maple Tree Identification Chart on page 26*

What to Do: Cut a 2-foot branch from a maple, and put it in a vase indoors; study its features every day. If it's winter, and you have only the buds, they may expand and grow into leaves when you bring the twig inside and put it into water. Every few days, you can measure how fast they grow. When you learn the name of your maple, children can make a label for it. Children can make their own **"Field Guide to Local Maples,"** showing in writing and pictures the identifying features of each species.

Name

What Kind of Pine?

Pine needles grow in bundles, in contrast to spruce and fir, whose needles grow singly. Counting the needles in each bundle is a fun and important step in identifying a pine.

- **Look at the pine branch and study its features.**
- **Then, fill in the answers to the questions asked in the chart below.**

How to Identify a Pine

How many needles are in each bundle?	
What color are the needles?	
Are they stiff, or flexible?	
Do they have a scent?	
Is the bark scaly or smooth?	
Are the cones prickly or sticky?	
Do the cones open up after a few days inside?	
What do the seeds look like?	

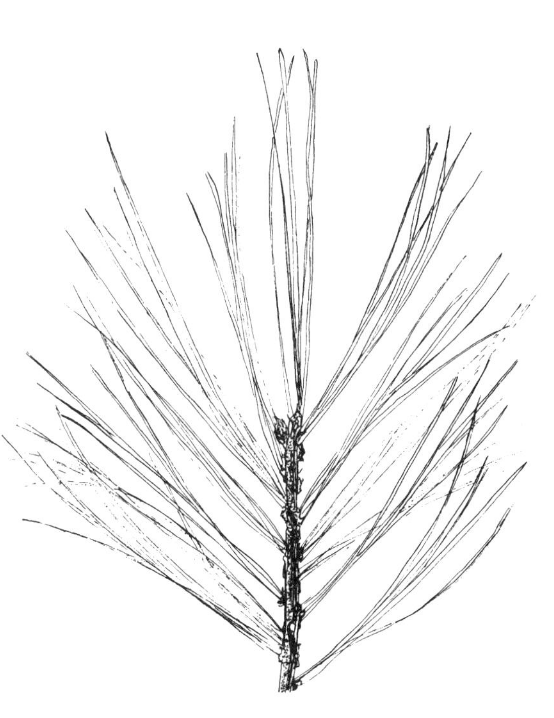

Draw a Picture of Your Pine:

Type of Pine: ____________________

Name

What Kind of Maple?

Maple leaves and twigs are opposite, not alternate like on most trees. Compare a maple leaf to your palm and hand. Do you see why the vein pattern is called "palmate"? Are there winged seeds on your maple tree? Study their shape, a great help in determining the species.

- **Look at the maple branch and study its features.**
- **Then, fill in the answers to the questions asked in the chart below.**

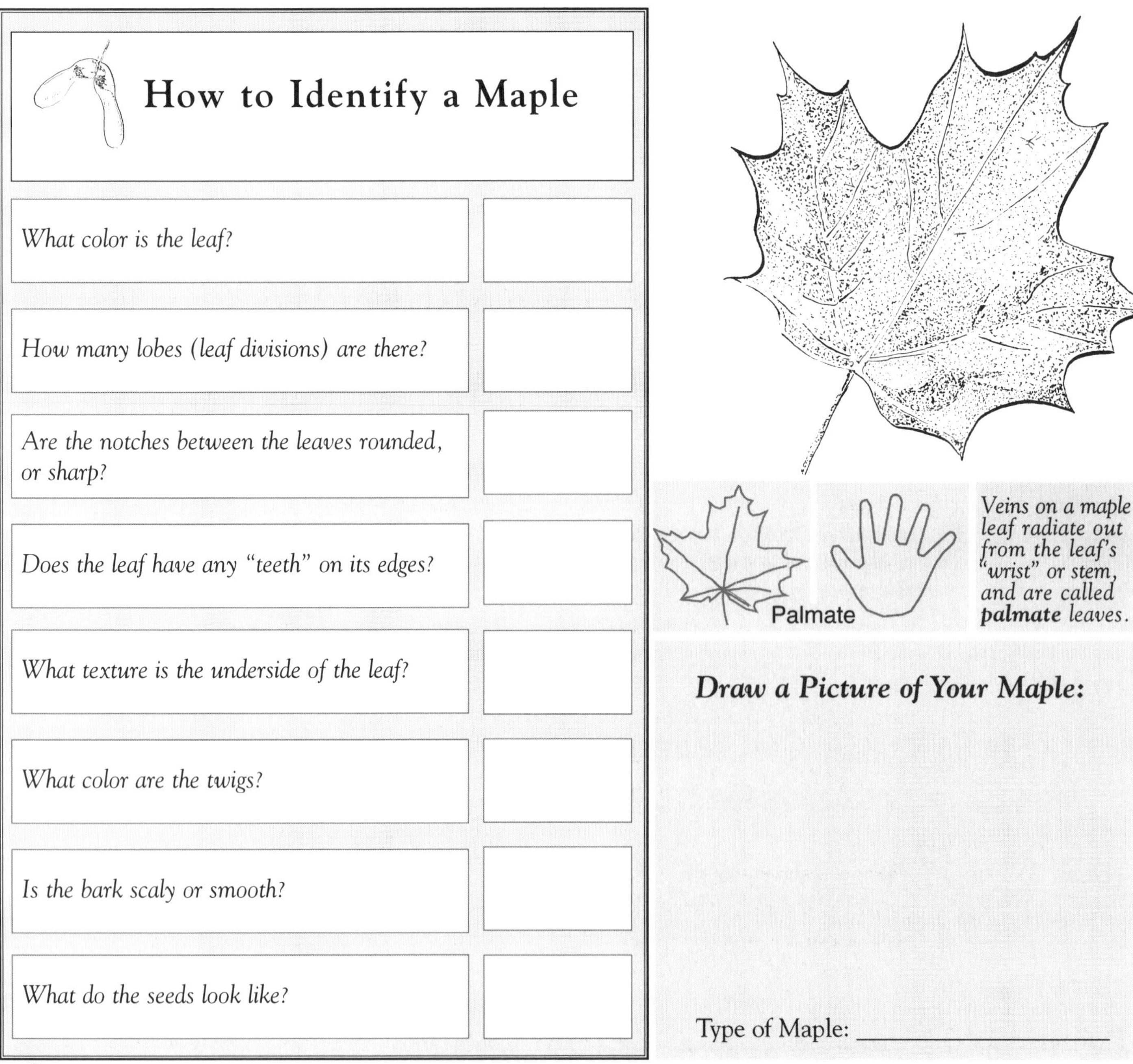

How to Identify a Maple	
What color is the leaf?	
How many lobes (leaf divisions) are there?	
Are the notches between the leaves rounded, or sharp?	
Does the leaf have any "teeth" on its edges?	
What texture is the underside of the leaf?	
What color are the twigs?	
Is the bark scaly or smooth?	
What do the seeds look like?	

Veins on a maple leaf radiate out from the leaf's "wrist" or stem, and are called ***palmate*** *leaves.*

Draw a Picture of Your Maple:

Type of Maple: ____________________

Activity 3: *Find an Oak*

Objective: To learn to identify an oak.

Concept: Oak trees of many species grow throughout the U. S. and Canada. Their seeds are *acorns*, a familiar treat for squirrels. Some students may know how to whistle by blowing on the edge of an acorn "cap."

You Will Need:
- *Clippers*
- *If necessary, permission to clip off a branch about 2 feet long*
- *A heavy, sturdy, non-tippy vase with water*
- *The Oak Identification Chart on Page 28*

What to Do: Cut a 2-foot branch from an oak, with acorns if possible, and put it in a vase indoors; study it daily.
Study the proportions of the leaves and acorns. If you have only the buds, notice how they are alternate (unlike maple) and clustered at the end of the branch.
Note all the differences you can between oak and maple.

Activity 4: *Find a Hackmatack*

Objective: To learn how to identify a hackmatack (also called tamarack or larch).

Concept: If you live in the Northeast or around the Great Lakes or in Canada, hackmatacks may grow in your area. They are also called larch and tamarack. Not so common as the other kinds of trees above, they are nevertheless very interesting. Hackmatacks are conifers, but are not evergreen! Their needles are deciduous—they turn yellow and fall off in the fall. Over the winter, hackmatacks look like dead skeletons of trees. The next spring, new, soft, chartreuse needles will emerge in bundles. Needles will slowly get stiffer, longer, and darker green. Their cones are purple at first, then turn the familiar, woody color of a mature cone. It's fascinating to watch them change over the seasons.

You Will Need:
- *Clippers*
- *If necessary, permission to clip off a branch about 2 feet long*
- *A heavy, sturdy, non-tippy vase with water*
- *The Hackmatack Identification Chart on Page 29*

What to Do: The hardest part might be finding a hackmatack! In the wild, they often grow in bogs. But sometimes they are planted in neighborhoods for variety. Clip off a branch and put it in a vase in water; see what changes you can observe. Better yet, if there is one in the neighborhood, visit it in every season and take notes on the changes.
How many differences between hackmatack and pine can you find?

Did you know? Tamarack comes from an Algonquin word, akemantak, meaning "wood used for snowshoes."

Name

What Kind of Oak?

The twigs and branches of oak trees are *alternate*. Oak leaves are usually indented *(lobed)* and have a vein pattern of a long midrib with side veins.

- **Look at the oak branch and study its features.**
- **Then, fill in the answers to the questions asked in the chart below.**

How to Identify an Oak

What color is the leaf?	
How many lobes (leaf divisions) are there?	
Are the lobes (if any) deeply cut?	
Are the tips of each lobe round, or sharp?	
What texture is the underside of the leaf?	
What color are the twigs?	
Is the bark furrowed or scaly?	
What do the acorns look like?	

Draw a Picture of Your Oak:

Type of Oak: ______________________

Name

What Kind of Hackmatack?

The pale green foliage of the hackmatack is extremely thin and delicate. This unusual conifer loses its leaves in the fall–unlike most conifers, which are evergreen.

- **Look at the hackmatack branch and study its features.**
- **Then, fill in the answers to the questions asked in the chart below.**

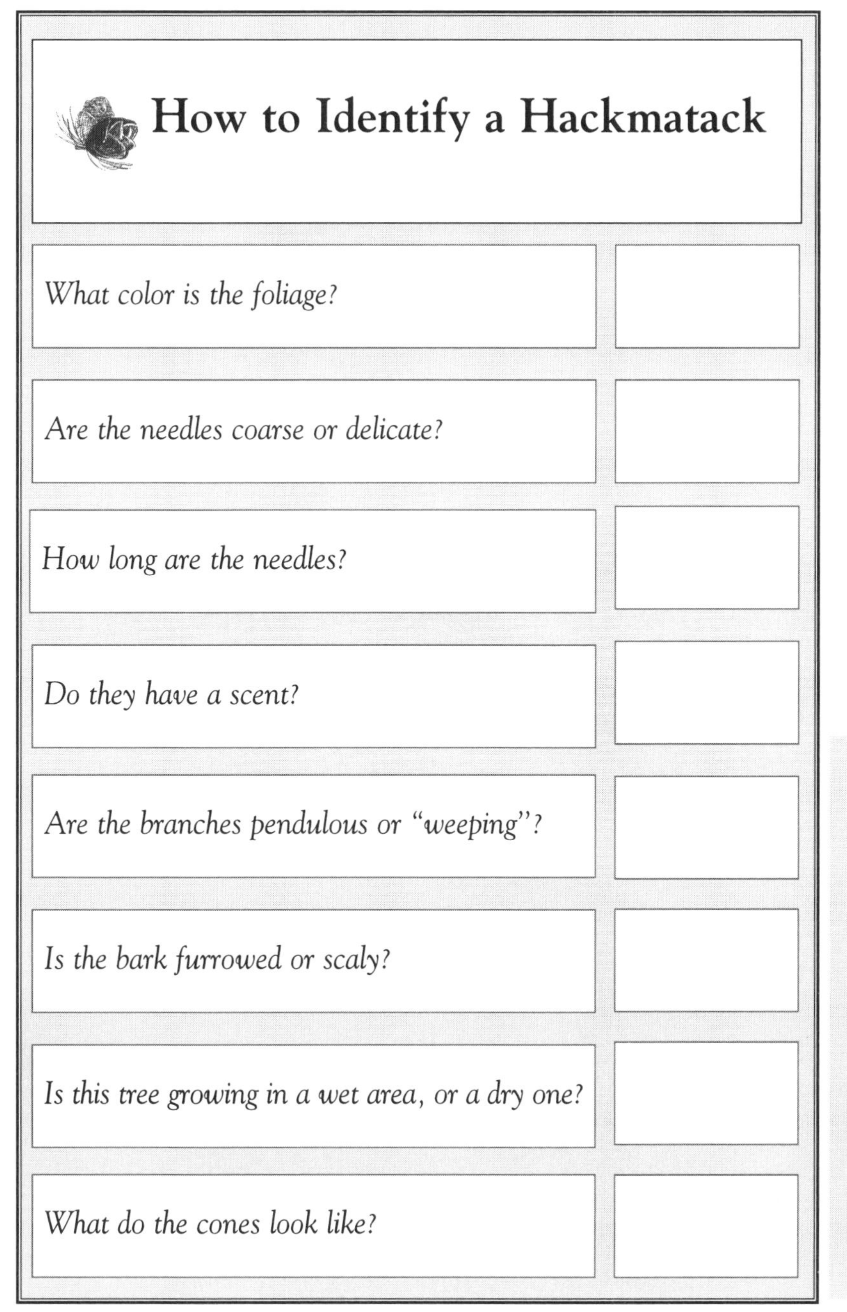

How to Identify a Hackmatack

What color is the foliage?	
Are the needles coarse or delicate?	
How long are the needles?	
Do they have a scent?	
Are the branches pendulous or "weeping"?	
Is the bark furrowed or scaly?	
Is this tree growing in a wet area, or a dry one?	
What do the cones look like?	

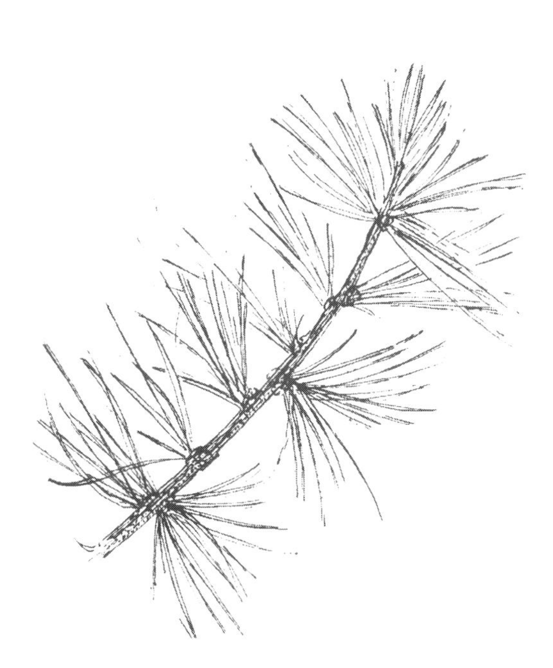

Draw a Picture of Your Hackmatack:

Type of Hackmatack: ____________________

Name

Leaf It to Me

When people think about trees, the first thing they think about are leaves. Leaves come in many shapes–oval, triangle, heart-shaped, feather-shaped, and hand-shaped are a few.

- **The leaves of the trees hidden in the Word Box all have different shapes. See if you can find and circle them in the puzzle below. The words may be spelled up and down, sideways, backward, or diagonally.**

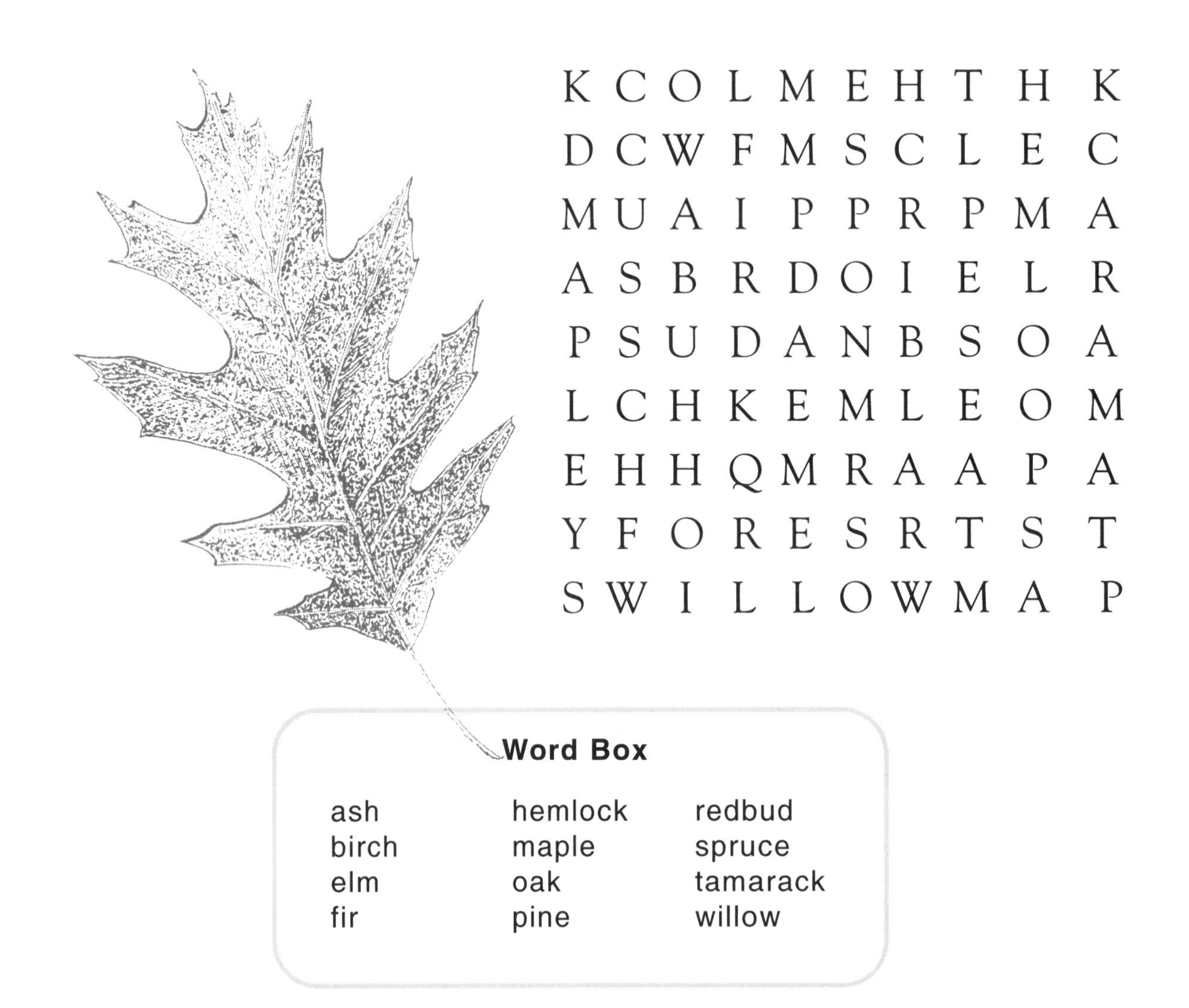

It's fun to collect leaves!

You can put them in a little bag or envelope until you get home. If you place a damp paper towel in the bag, the leaves won't dry up or curl. Then, carefully place them between a few sheets of newspaper, and put some heavy books on top. In two weeks, they'll be completely pressed and ready to display in your album or stored in a box.

CHAPTER FOUR
Layers in the Forest

The complex structure of a mature forest means that there are many different places for animals to live and hide.

Here's another way to look at a forest: learning to see its layers. This takes a little practice—at first, all you see is the chaos of many trees crowded in together. These are the layers usually found in a mature forest:

- the canopy (the upper leaves and twigs that live in the sun)
- the understory (the shaded layer of lower branches and leaves)
- the shrub layer
- the layer of herbaceous (non-woody) plants
- the soil (which in turn has layers of its own)

Students can think of the forest as a house: the canopy is like the roof; the soil is like the basement; and there are layers like floors in between.

> *Cedar and pine, and fir, and branching palm,*
> *A sylvan scene, and as the ranks ascend*
> *Shade above shade, a woody theatre*
> *Of stately view.*
>
> -Milton

Different animals live in the different layers. Some birds, like orioles, are canopy species, foraging in the sun, eating insects off the upper leaves. Some birds, like chickadees, tend to live in the darker understory. Woodpeckers live, forage, and make their homes on tree trunks, not on the outer twigs. Some birds live mostly in the shrub layer, or look for insects under the leaves on the forest floor. The same is true of frogs: there are frogs of the leaf litter and there are tree frogs. Squirrels are similar in that some species live in the trees and some (like chipmunks) make their homes in burrows under the soil.

The leaf litter has many small animals living in it: insects, millipedes and centipedes, sowbugs, springtails, slugs and other snails, and the familiar earthworm. Most of these small animals eat and digest the leaf litter, helping turn it into soil. (The centipede, however, is a predator of the others.)

Smaller salamanders (like the red-backed salamander) live in the layer between the leaf litter and the topsoil. They eat the smaller animals mentioned above. Larger salamanders (like the spotted salamanders) actually make tunnels in the topsoil, and eat worms, larger insects, and spiders.

Shrews, the smallest mammals, are tiny, fierce predators of the leaf litter, eating anything they can catch as they charge frenetically among the fallen leaves.

Moles also tunnel in the soil, and so do earthworms. All this tunneling aerates the soil, and also enables the rain to soak into the soil more easily. All this tunneling, plus the leaf litter and mosses, make the forest floor a soft and delightful layer to walk upon!

Name

Forest Minibeasts

Many of the little creatures that live in the leaf litter are not insects! Insects have six legs. Millipedes, centipedes, daddy longlegs, and sowbugs are related to insects, but have more legs. Other small animals of the forest floor are slugs (which are a kind of snail) and worms.

•Below are some facts about these tiny animals of the forest. Circle the facts that are new to you.

Which one is faster – the millipede or centipede? The millipede is actually very slow despite its many legs; the centipede moves quickly.

A Daddy longleg's legs are very delicate and fall off easily. Some Daddy longlegs' legs will twitch for a minute, others for an hour after they are detached. This helps to distract predators while they escape.

A millipede has hard plates on its back for protection, and it can curl up if attacked. It also produces a nasty smelling liquid to disturb its enemies.

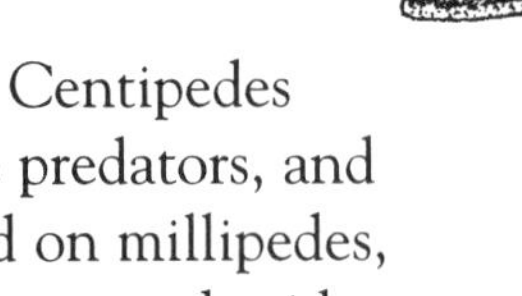

Centipedes are predators, and feed on millipedes, sowbugs, and spiders. Centipedes rarely bite humans.

Slugs slide along on a layer of their own mucous.

Slugs have a file-like tongue with 27,000 teeth on it. They use their tongue to scrape food off trees, other plants, and rocks.

A female sowbug holds her 25-75 babies in her pouch for up to a month.They look exactly like mom except for their tiny size!

Centipedes don't always have 100 legs; they can have anywhere from 20 to 200.

Slugs have antennae which they can extend or contract. They also have a breathing hole on their side, which they can open or shut.

Activity 1: Make a Mural of Layers of the Forest

Objective: Children will learn to look at a forest and interpret it in terms of its layers.

Concept: The layers that form in a mature forest give animals many different places to hide and to find food. The upper layers get the most sunlight, while it is progressively darker in the lower layers. Many animals have an affinity for certain layers more than others.

You will need:
- *Large sheets of newsprint or other paper*
- *Art materials*

What to Do: Have the students research what common local animals are found in which forest layers; make a mural with forest layers depicted.
They can make more or less life-sized images of the animals, and attach them to the mural in a suitable spot. They can put in *people* as part of the ecology of the forest, too: perhaps foresters or loggers, field biologists, birdwatchers, hikers, or other people enjoying recreation in the forest.

Which layer?

caterpillars • birds • squirrels • deer • butterflies • beavers • porcupines • spiders • skunks • lizards • rabbits • blue jays • bluebirds • snakes • bats • treefrogs • wood ducks • beetles • flying squirrels • salamanders • mice • moose • opossums • wild turkeys • ovenbirds • woodpeckers • turtles • worms • bears • orioles • foxes • centipedes • owls • bobcat • chipmunks • daddy-longlegs • woodland-caribou • toads • mosquitoes • red-squirrels • lynx • coyotes • shrews • puma • hares • millipedes • ermine • martens • hawks • insects • weasels • raccoons •

Name ____

Crack the Code

Learning about the forest and its inhabitants can be a lot of fun.

- **Try to find out the answer to this Forest Riddle.**
- **Use the code to fill in the letters. The first letter has been done for you.**

Riddle: If a mouse loses its tail, where can it get a new one?

Answer:

A ___ ___ ___ ___ ___ ___ ___ ___

___ ___ ___ ___ ___

Clue Box

A =

I = 

O = 

L =

E =

T = 

S =

R = 

Mousey, mousey

Mice are an important part of the forest in many ways. They hop around in the forest, climbing shrubs, gathering seeds and berries, storing them under a log or in an old bird's nest. Some of these seeds will grow into a new shrub or tree. So mice are an important planter of seeds! Also, mice are an important food for foxes and owls.

CHAPTER FIVE

Sun and Shade in the Forest

It's often dark and shady in the forest. Usually the air is cooler and more moist than the air outside the forest. Sometimes only a few sun flecks reach the forest floor. It can sometimes be difficult for new young trees to grow in the deep shade underneath a "closed-canopy forest," one in which the canopy closes off the light to the forest floor.

All trees need sunlight to grow. Sunlight gives them the energy they need to do their miraculous work: to convert air, water, and a few minerals into sweet sap, oxygen, and wood. But growing in the full sun presents its challenges, too. The forest soil can get dried out, desiccating the new seedlings. A hard rain can then cause erosion of the forest soil.

In the small-scale woodlot forestry in *Shelterwood*, the grandfather did not cut all his trees "like hay," but left some trees to shelter the new seedlings growing on the forest floor.

> ***The woods are lovely, dark and deep.***
>
> ***–Robert Frost***

Different tree species grow best under different conditions.

Some trees cannot grow in the darkness of the forest floor. They don't get enough sun. Their seeds may fall in the forest, but never get to grow. From sun-loving trees, only those seeds that fall in the sunlight will grow into trees. Some kinds of birch, poplar, and cherry are examples of trees that need lots of sun to grow. Only in a clearing in the forest, or outside the forest, will they get their day in the sun, and a chance to grow.

You may have seen a field that was "abandoned." That is, it was no longer mowed, or cattle no longer grazed it down, and the field got overgrown with small weedy trees. Over time, the small scrawny trees will grow, and the forest will reclaim a field. The first trees to grow in such a field will be the sun-loving species. The wind will bring in seeds of birch and poplar and white pine, and birds will drop cherry seeds. These trees will grow fast in the sun.

Often, in the shade of these fast-growing trees, you can find seedlings of trees with a different strategy of growth. These trees can grow with less sunlight, under the darker conditions of the understory. Their seeds can get a start, and grow slowly, under shady conditions. Beech and hemlock are good examples.

Sometimes you can find a patch of forest in which one species dominates the canopy, and a different species dominates the understory. The understory species is more shade tolerant than the canopy species. Such a patch of forest is a good place for students to study forest succession, an ecological term for the changes that occur as a field grows into forest, or as a forest matures. Generally, in early stages of succession, sun-loving, short-lived trees will dominate. As a forest matures, shade-tolerant, long-lived trees will slowly overtop, shade out, and take over from the sun-loving but short-lived early successional trees.

Activity 1: How Dark Is the Shade?

Objective: Children will learn that the amount of shade and sunlight are important factors in a forest. They will learn a simple way to measure them.

Concept: Some forests are darker than others. Some trees (like hemlock and beech) have very dense foliage, making for a very dark forest floor. Some trees (like many kinds of birch and poplar) let lots of sun through their foliage.

You Will Need:
- *A sunny day*
- *Clipboards*
- *White paper with one side blank (a good reuse of used office paper); or, the Forest Shade Worksheet on page 37*
- *Trees or woods nearby*

What to Do: Have the children make a grid on the blank side. A grid of 1-inch squares, 8 x 10 inches, works well. Then, stand in the forest or under a tree and compare the number of mostly sunlit squares and mostly shaded squares. This number enables you to compare the shadiness of different places in the forest, or under different trees.

For the most part the woods are dark and silent,
but here and there one comes out into open areas
of sunshine and woods smells.
–Rachel Carson

Activity 2: Find a Sign of Change in the Forest

Objective: Children will learn to think of the forest as dynamic, not static. (In forestry, the study of forest change over time is called "forest succession" or "forest dynamics.")

Concept: When most people look at a mature forest casually, they assume it has always been there. They think of big trees as permanent entities, but in reality the forest is always changing. Everywhere you look, you can see signs of change. With practice, you can learn to think of the forest as ever-changing, not static.

You Will Need:
- *Just yourself, and a group of children, and a willingness to think of the forest in a new way*

What to Do: Tell the children that, even though the forest seems eternal, it is always changing. You don't always see the changes happen, but you can see the evidence of change. Show them examples: a twig on the ground used to be a branch on a tree; a brown leaf on the ground used to be a green leaf on the tree; a small tree used to be a seedling, which used to be a seed; a log on the ground was once a standing tree. Let them find and interpret many examples of their own.

Name

Forest Shade Worksheet

Here's an easy way to measure the degree of canopy cover in an area of forest. Stand in the forest under a tree or group of trees and hold your worksheet flat.

- Color in the squares that are shaded.
- Count the number of squares shaded and refer to the table below.

How many squares are shaded? Write your answer here: _______ squares.

Circle the box that correctly estimates how shady it is under your tree or in your area of the forest.

Number of shaded squares:	Under 17: Light Shade	17 - 34: Moderate Shade	Over 34: Heavy Shade

Activity 4: A Shady Look at the Forest

(For older children who already know their trees, and can understand the long time frame of change in the forest.)

Objective: Children will learn to observe the forest from the point of view of forest succession.

Concept: As the forest matures over a long time, layers develop, including a canopy and an understory. Trees in the understory will usually be slow-maturing, shade-tolerant types. There are other variables, such as tree longevity, that affect patterns of succession. But educationally, it's best if students deal with one variable at a time.

You Will Need:
- *A study site in which the canopy is* ***clearly dominated by one species*** *and the understory is clearly dominated by a* ***different*** *species*
- *A list of local forest trees and their level of shade tolerance*

(A good study site is very important in this activity so that children don't become confused by a site that presents too much complexity.)

NOTE: *Here is a simplified list; perhaps a local forester can give you additional information about your local trees.*

Degree of Shade Tolerance of Species

Western Trees:

Species	Tolerance
ponderosa pine	intolerant
Douglas-fir	intolerant
western hemlock	tolerant
white fir	tolerant

Eastern Trees:

Species	Tolerance
aspen	intolerant
longleaf pine	intolerant
slash pine	intolerant
white pine	middle
red spruce	tolerant
red maple	tolerant
sugar maple	tolerant
balsam fir	tolerant
hemlock	tolerant

What to Do: Have the children make a diagram of a forest with a canopy layer and an understory layer.
Identify trees in both layers and try to figure out whether there is a pattern of shade-tolerant trees in the understory, and sun-loving species in the canopy.

Note: If the canopy as well as the understory both consist of shade-tolerant species, that may be due to the maturity of the forest, or possibly, forest management practices.

CHAPTER SIX

Wind, Water, and Soil in the Forest

The previous chapter, "Sun and Shade in the Forest," presented how older forest trees can act as a "shelterwood" for younger trees. The grandfather in *Shelterwood* did not cut all his trees, but left some to help protect seedlings against the searing light, heat, and drought of a long day in the sun. Trees can act as "shelterwood" against the effects of wind and rain, as well.

Trees are good protection against the wind. In windy, open areas, people plant "windbreaks" or "shelterbelts" of trees to protect their homes from the wind. The forest interior is not typically a windy environment, because the trees stop the wind.

Wind is a fascinating thing as it relates to trees. Have you ever seen a tree that was blown down by the wind? A tree trunk with its branches on the ground and its roots up in the air is a dramatic and memorable sight!

The author John Muir, writing about the forests of California, said that "trees travel, out and back again" when they wave in a high wind. He clung to a treetop to experience a storm. He said the sound of a storm in the forest was like a symphony, with different kinds of trees making different noises in the wind.

Who has seen the wind?
Neither I nor you;
But when the leaves hang trembling,
The wind is passing though.

Who has seen the wind?
Neither you nor I;
But when the trees bow down their heads,
The wind is passing by.

–Christina G. Rossetti

Rain can be a dramatic and exciting thing to observe, too. While trees need rain, a hard rain can be harsh for tiny tree seedlings fully exposed to the elements. A hard rain can dislodge and erode soil. Water can pool around the base of a tiny tree, drowning it. If you stand outside in the open, in a heavy rainstorm, you will soon be soaked. If you stand under a tree with dense foliage, you will be somewhat protected. In a warm rain, it's fun to compare the difference between standing in the open and standing under a tree. Ask children if they have experienced the difference.

Some of the rain that lands on a tree is delayed in reaching the ground as it drips through the foliage. Some of the water evaporates before it reaches the ground. Some of the water runs down the tree's trunk. You can watch this happen from inside in a hard rain.

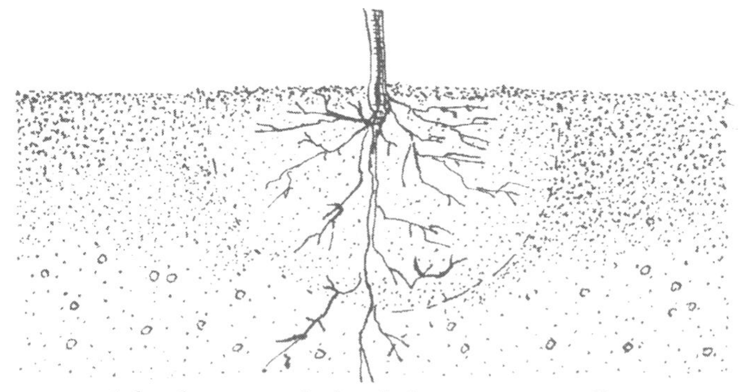

A hard rain can be harsh for tiny tree seedlings fully exposed to the elements.

In a garden a heavy rain will hit the bare soil hard and splash mud around, but in the forest the rain falls gently under the shelter of the trees. The soil is further protected by a layer of fallen leaves, twigs, and needles.

Children have experienced how trees protect them—from the sun, wind, and rain. They have experienced "shelterwood." They will enjoy experimenting with and measuring these protective qualities of trees. Children also love investigating forest soil. After all, they can have fun, get good and dirty, and call it science!

Because children have experienced wind, rain, and soil, they love related hands-on science activities. Furthermore, reading, writing, math, measuring, and making tables and graphs are a lot more interesting when they are based on vividly memorable, firsthand experiences.

Nearly everyone has experienced "shelterwood," the protective presence of mature trees. In the book *Shelterwood*, these trees symbolize the caring protection that parents seek to provide for their offspring. The "shelterwood" trees also symbolize the caring protection the grandfather provides for Sophie as well as for his forest.

Activity 1: Measure the Wind, Inside and Outside the Forest

Objective: Children will learn simple ways to measure the wind, and will learn whether it's windier inside the forest or out in the open.

Concept: Trees do a good job of blocking the wind–especially many trees together. That's why people plant trees as a windbreak, and why there is much less wind in the forest than in an open area with nothing to block the wind.

You Will Need:
- *A nice, breezy day*
- *A garden stake with a ribbon attached to the top*
- *Additionally, a fun-to-use wind gauge can be bought from a science supply house for under $20*
- *Some home and garden stores also sell inexpensive wind gauges*

What to Do: Go outside and into an open area, like an athletic field, unsheltered by buildings, trees, etc. Insert the stake into the ground, stand back, and observe the ribbon. *Does it move? How often and how much?* You can time the movement, or estimate how far from the stake the ribbon moves. Then, try the same thing 300 feet into the woods. (Scientists estimate that the "edge effect" of increased wind and light may go that far into a forest.) *How can you decide which environment is windier?*

Activity 2: Measure the Wind Another Way, Inside and Outside the Forest

Objective: Children will learn a simple way to measure the wind, and will learn whether it's windier inside the forest, or out in the open.

Concept: It is fairly easy to estimate wind speed, even from inside, by carefully observing what's happening around you and using a Beaufort scale. You can use the Beaufort code number, saying, for example, "It looks like Beaufort four, a moderate breeze, about 15 miles per hour."

You Will Need:
- *A nice, breezy day, to get practice*
- *The simplified version of the Beaufort scale–on page 41–for each child*

What to Do: On a nice, breezy day, try using the Beaufort scale in an open, unprotected area, then 300 feet into a forest. (Scientists estimate that the "edge effect" of increased wind and light may go that far into a forest.) *Is there a difference?*
A safety note: Do not go into a forest on a very windy day, when branches and even trees may fall. On a very windy day, children can enjoy using the Beaufort scale safely from inside, by watching through a window.

Name

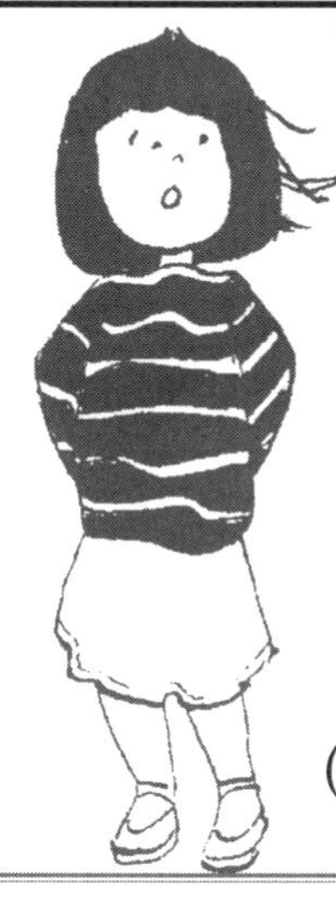

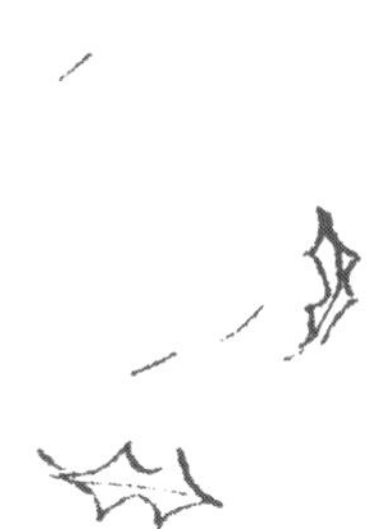

Measure the Wind
with the

Beaufort Scale

(Adapted and Simplified for Children's Use)

Code	Description	Signs	Approximate wind speed in miles per hour
0	Calm	Calm; smoke rises vertically	0
1	Light Air	Smoke drifts; weathervane does not move	2
2	Light Breeze	Wind felt on face; leaves rustle weathervane moves	5
3	Gentle Breeze	Light flags blow; leaves and small twigs move constantly	10
4	Moderate Breeze	Small branches move; papers blow; dust is raised	15
5	Fresh Breeze	Small trees sway; crests form on inland waterways	20
6	Strong Breeze	Overhead wires whistle; large branches move; umbrella used with difficulty	28
7	High Wind or Moderate Gale	Whole trees in motion; walking is difficult	35
8	Fresh Gale	Twigs break; progress of people is impeded	42
9	Strong Gale	Insecurely fastened parts of houses tear loose	50
10	Whole Gale	Trees are uprooted; buildings are damaged	60
11	Storm	Widespread damage	70
12-17	Hurricane	Devastation occurs	72+

Activity 3: **Use a Rain Gauge**

Objective: Children will learn how to measure and compare rainfall.

Concept: Rainfall is an important part of the environment. Garden stores sell simple rain gauges, graduated tubes that catch and measure the rainfall.

You Will Need: • *A rain gauge, typically available for under $5 from a garden supply store*

What to Do: Children can set up one rain gauge under a tree or inside a forest, and one out in the open. The location should be selected by an adult with safety in mind—you don't want anybody tripping or falling on the rain gauge, or mowing over it. Children should check the gauge after every rainfall, recording the date and comparing the amount systematically in both locations, then emptying out the water. *Where does more rain fall—under the trees or out in the open?*

Activity 4: **Splash Board**

Objective: Children will learn how to observe and measure how vegetative cover protects the soil from erosion.

Concept: Rain falling hard in the open, on bare ground, can remove topsoil and start the process of erosion. Trees and other plants protect the soil.

You Will Need: • *Two plywood squares, about 2 x 2 feet each, painted white*

What to Do: Children can dig two narrow 2-foot trenches in the ground, one in the open and one under a tree or in the forest. The location should be selected by an adult with safety in mind—you don't want anybody tripping or falling on the splash board. Bury the edge of the board about 6 inches in the ground so the white board stays upright. Check the splash boards after a heavy rain. *Is there evidence of mud splashing on the board? Which one is more mud splashed? Which site is more likely to erode?*

What About Worms?

- Worms eat their way through the soil, helping air and water reach deeper into the ground.
- A worm's body is covered with tiny hairy bristles.
- Their tunnels may be 5 feet deep.
- The thick band around the worm is where they store their eggs. The belt moves down to the end of the worm and falls off, forming a protective cocoon around the eggs.
- Worms have light-sensitive receptors all over their bodies that help them find their way around in their environment.

Which way is which?

- The more pointed end is the worm's head.

Activity 5: Soil Layers Outside

Objective: Children will learn to see the subtleties of soil color, and perhaps see some of its layers.

Concept: Where the soil has been undisturbed, i.e., not plowed, bulldozed, or trucked in to make a lawn, it sometimes occurs in layers of different chemical composition. These layers sometimes have subtle color differences which take some practice to see. When children first compare soil, it all looks the same to them. Gradually, with practice, they can see the subtleties of soil color—perhaps yellowish-brown, orange-brown, or gray-brown. Usually there will be a top layer of organic matter—fallen leaves or needles in the forest, or a mat of dead grass in a field or on a lawn. Around a schoolyard, it can be challenging to find areas of undisturbed soil, but usually you can find places with different colors of soil.

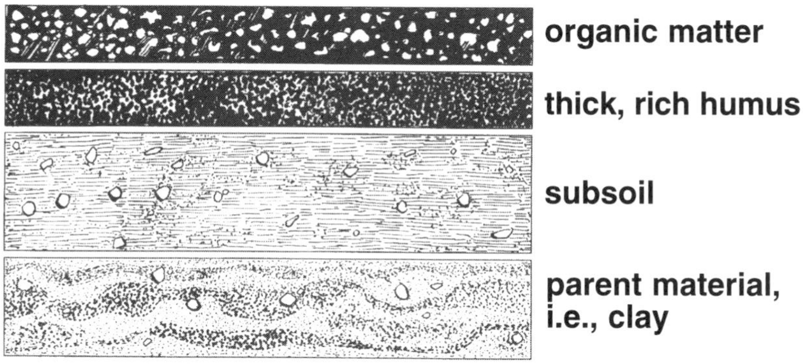

You Will Need:
- *A spade*
- *A spirit of exploration*

What to Do: Dig a hole at least a foot deep, noting what the organic layer on top is made of, and examining the soil for color. It helps to roll a marble-sized ball of moist soil in the hands, and compare it in good light to a small "soil-ball" from a different spot. That will help children see the color differences. *What do the color differences tell us about the soil?*

A safety note: Be sure children refill the hole, so nobody falls in it.

organic matter
thick, rich humus
subsoil
parent material, i.e., clay

A Soil Profile

Activity 6: Soil Layers in the Classroom

Objective: Children will learn a different way that soil layers are sometimes created—not through soil chemistry, but through the action of water. Children will gain practice in seeing subtle differences in soil color and texture.

Concept: When soil mixes with moving water, and then the water stands still, the heaviest soil particles settle out first, and the lightest particles settle out last. This process often leaves different layers of soil.

You Will Need:
- *A large, clear plastic jar with top (Plastic jars are safer for children to handle than glass jars)*
- *Enough soil to fill the jar one-third full*

What to Do: Put the soil in the jar, then fill it nearly full with water. Shake it up. Put it on a shelf in view of the children and let it settle all day, overnight, or perhaps for several days. Children should record how the contents of the jar change over time. *Why are layers formed in the jar?*

Activity 7: **Soil Scientist**

Objective: Children will learn a simple, fun soil test for clay or sand content of soil.

Concept: Soil scientists have a simple soil test that children love—making a "soil snake" by taking moist soil and rolling it in the hands. Soil with some clay content will make a fine snake, but sandy soil just will not make a snake, no matter how moist it is and how much you roll it. Clay has very tiny particles that stick together, while sand particles are larger and do not stick together. Soil with high sand content also feels gritty.

You Will Need: •*A spade and, again, a spirit of exploration*

What to Do: Try to find two places nearby with moist soil of contrasting texture, color, and moisture levels. Have children try making "soil snakes" and comparing results with different kinds of soils. *How can they explain different results? Can they now make a table showing soil sample locations, and soil differences in color and texture?*

Activity 8: **How Wet Is this Site?**

(For older children who already know how to identify trees.)

Objective: Children will learn to predict how wet the soil is by the kinds of trees growing there.

Concept: Some tree species can grow where the soil is wet. **Examples: cedar, hackmatack, black spruce, willow, red maple, silver maple, brown ash, and baldcypress.** Other tree species grow best in deep, well-drained soil. **Examples: white pines, loblolly pines, ponderosa pines, Eastern and Western hemlock, Douglas-fir, red spruce, white spruce, white ash, American beech, and sugar maple.**
(Keep in mind that the season and recent weather will affect your moisture findings.)

You will need:
- *Two contrasting forested sites, one wetter, the other drier*
- *A spade*
- *A list of local trees that grow in wetter soil and in drier soil*
  You can use the above, simplified list, or a more complete (but sometimes confusing) list from a local source
- *Each student will want a notebook or clipboard to record findings*

What to Do: Before going out to the two sites, students can develop their own soil wetness scale, such as: very wet, wet, medium, moist, dry. Or, you can lead them into creating a vocabulary-expanding scale: inundated, saturated, wet, moist, dry, desiccated. Then, take students to each site. Look at the tree species and try to predict how wet the soil will be. Have them dig a hole and explore the soil. *How wet is it?* Wetter soil will certainly feel and look wetter. It will often smell different from upland soil, making a memorable comparison. It will often be darker-colored and grayer than upland soil, which often will have relatively brighter colors: yellowish or orange tints. (Remember it takes practice, and good light, to see differences in soil color.)
For safety, be sure you fill in your soil pit after you're done with it.

Name

Soil Scientist Worksheet

for Activities 7 and 8 on page 44

Activity 7: **Soil Scientist**

- Make a "soil snake" by taking moist soil and rolling it in your hands. Choose samples with contrasting texture, color, and moisture levels.
- Compare your results in the data table below.

Soil Type Test

Soil Type #	Soil Sample Location	Soil Texture	Soil Color	Amount of Moisture
1				
2				

Activity 8: **How Wet Is this Site?**

- Look at the tree species on each of the two sites chosen and see if you can predict how wet the soil will be.
- Write the number that corresponds to the wetness of your soil samples in the blank spaces below. How did your predictions compare with the results?

Soil Wetness Scale

1 Inundated **2** Saturated **3** Wet **4** Moist **5** Dry **6** Desiccated

Location	Types of Trees on Site	Is color of soil darker, or grayer?	Yellowish or orange tints?	Predicted Wetness of Site: #	Actual Wetness of Site: #
Wetter Site					
Drier Site					

Tree Species That Grow In Wetter Soil	
Cedar	Red maple
Hackmatack	Silver maple
Black spruce	Brown ash
Willow	Baldcypress

Tree Species That Grow In Drier Soil		
White pine	Western hemlock	White ash
Loblolly pine	Douglas-fir	American beech
Ponderosa pine	Red spruce	Sugar maple
Eastern Hemlock	White spruce	

Name

Litter Bugs

Animals that live in the leaf litter and the first few centimeters of soil are a big help to the bacteria and fungi in breaking down the leaves and dead wood on the forest floor. These nutrients can then be used by plants to make new leaves and wood.

- **Circle the names of the leaf litter animals hidden in the puzzle.**
- **The words may be spelled up and down, sideways, backward, or diagonally.**

Word Box

springtail
spider
millipede
wood louse
daddy-longlegs
mite
slug
centipede

D A D D Y L O N G L E G S
W E L L Q J P N Y S L E R
O E B E P A U D P Q O U S
O P D D W O O R L Z L E G
D E Y E S P I D E R L O N
L J D P P N L Q S R E D E
O S H I G I D K B Z R T P
U M I T E S L U G P E D E
S T A N O V A L H Q D A D
E I U E Q F Z D I V D Y L
L N S C G M P D H M M I T
S P R L O G W O L S R E D
M I L S P E T I G U L O D

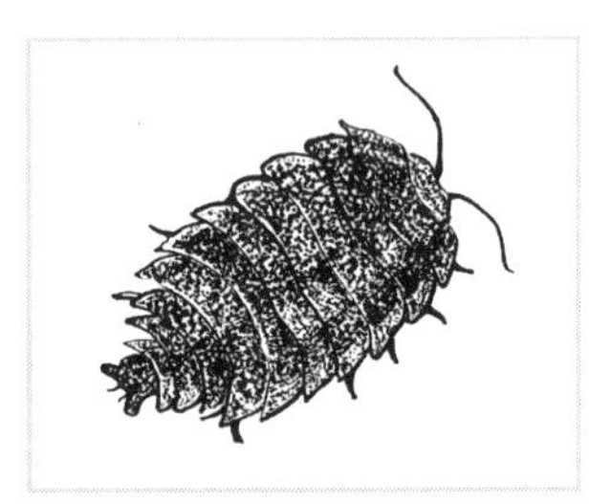

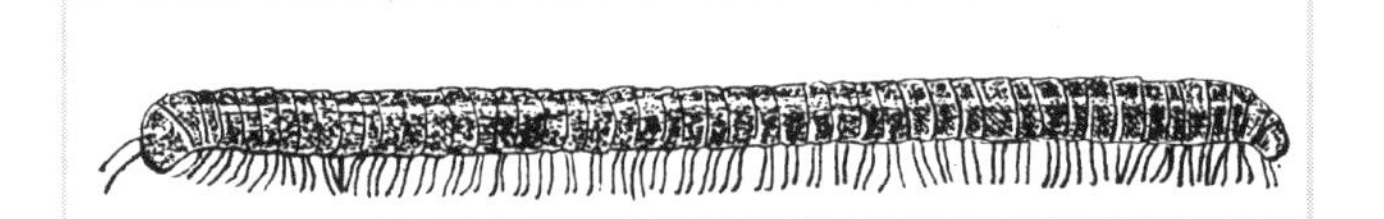

Did you know?

The sowbug is not an insect. It is a land crustacean related to the lobster.

Millipedes may have up to 400 legs, but not 1,000 as the name implies.

Love those Leaves!

Leaf litter animals play a very important role in the forest food web.
For instance, a mite will eat dead leaves and fungus, this mite will be eaten by a spider, the spider will be eaten by a beetle, the beetle will be eaten by a mouse, and the mouse eaten by an owl!
Talk about recycling!

CHAPTER SEVEN

Baby Trees, Big Trees, Dead Trees

"Great oaks from little acorns grow." The saying is true, but it's hard to see a tree actually grow. You may have seen a tree sprout from a seed over a summer, or you may have planted a seedling and watched it grow slowly into a small tree, called a sapling.

Most children have not had that experience. They usually think of trees as big, old, and unchanging. To understand forestry, they need to understand tree growth and perceive trees and forests as dynamic and changing entities.

> *"Oak trees come out of acorns, no matter how unlikely that seems. An acorn is just a tree's way back into the ground."*
>
> *- Shirley Ann Grau*

They will enjoy seeing a seedling and a young tree. Sometimes, near a big oak or a maple, you can find new seedlings with the characteristic oak or maple leaves (see Chapter 3). Sometimes a nursery or a forestry company will give you a seedling for educational use. It's best to get a seedling of a kind of tree the children are familiar with, as they can compare the seedling's leaves with those of the mature tree. Children may then plant the tree. They will need much patience, over several years, to watch it grow.

Sometimes, in the yard or schoolyard, you can find evidence of a tree's growth, perhaps where a tree has grown around or through wire fencing.

More evidence of how trees grow can be seen in a slice from the trunk of a tree that has been cut down. Children love learning how to tell the age of a tree by counting the annual growth rings from the tiniest one in the middle to the outermost, last growth ring. Species with distinct, easy-to-see growth rings include red (Norway) pine, spruce, and ash. People who cut a lot of wood (firewood users, carpenters, loggers) are often willing to give you some tree cross-sections for classroom use, so children can see the growth rings.

Patterns in the width of growth rings tell a story. It's fun to wonder and guess about the story. Most tree slices show regular, continuous growth. Most don't tell such an interesting story as these!

Tell your tree-slice supplier that your children have learned the basics of counting tree growth rings, and now are ready to learn how to interpret unusual patterns of growth. Your supplier likely knows how to find a tree with an interesting pattern, and needs only to be asked.

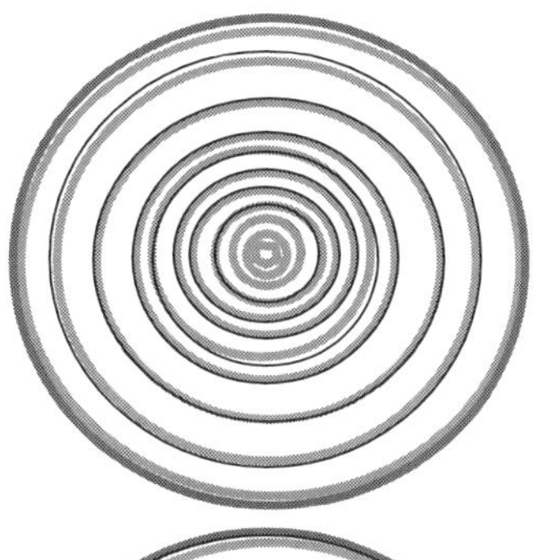

***Pattern A** tells you that the tree grew slowly for about 8 years, and then suddenly started growing faster, because the growth rings are wider. Perhaps the tree's neighbor-tree fell down and this tree got more sun and water after its 8th year.*

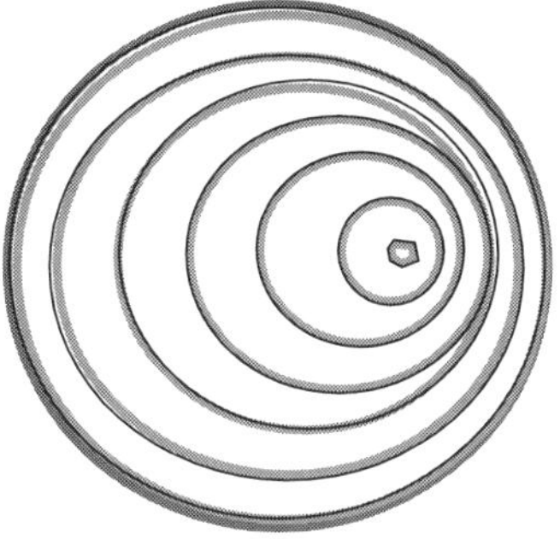

***Pattern B,** in which the growth rings are off-center, tells you that this tree was a "leaner." A leaning tree grows asymmetrically, typically putting on more wood on the lower side, for support.*

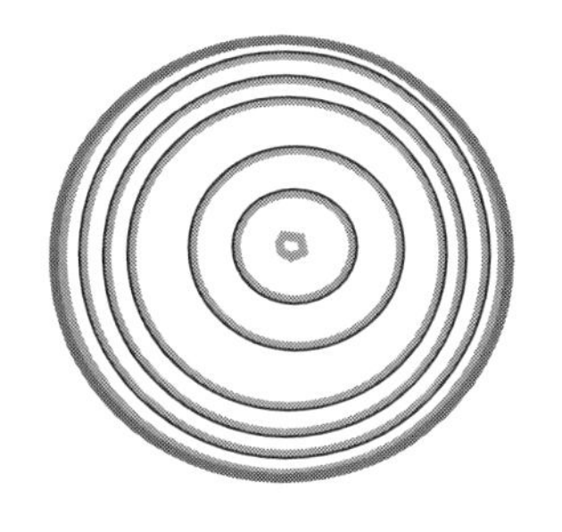

***Pattern C** tells you that the tree grew fast in its earlier years, 1 through 4. Then suddenly it didn't grow as fast for the next three years. Perhaps its neighbor tree grew bigger and was shading it at that time. Or perhaps there was a drought in those years.*

Tree growth is fascinating–but it usually happens v-e-r-y s-l-o-w-l-y. And trees die slowly in the forest, too. It's a long process. Understory trees may be crowded by other trees, and sometimes get "shaded out" and weaken. Then they are more vulnerable to diseases and insects.

Dead trees are an important part of the forest, too. While it is still standing, a dead tree is called a "snag" and is home for many species: fungi, insects, woodpeckers, squirrels, bears. After many years, it will fall over, becoming a log on the forest floor, sheltering more species (such as salamanders) as it turns slowly into soil.

Activity 1: Study a Tree Cross-Section

Objective: Children will learn about the growth rate of a tree.

Concept: In the forests of the U. S. and Canada, trees grow rapidly in the spring and early summer, and more slowly in the late summer and fall. This difference creates annual rings in the wood, which you can easily see in a slice from a tree trunk.

You Will Need:
- *Rulers*
- *Several tree cross-sections*

People who cut a lot of wood (firewood users, carpenters, loggers) are often willing to give you some tree slices for classroom use. Softwood tree rings are sometimes more easily seen than the annual rings on some hardwoods. Species with distinct, easy-to-see growth rings include red (Norway) pine, spruce, and ash.

What to Do: First, children need to learn how to count the rings. It is easiest to start in the center and count outward. Also, it takes some practice to interpret what is a growth ring and what isn't. It's best to explain that a count of the rings is considered an estimate of the tree's age. Two people counting rings on the same cross-section might come up with somewhat different ages. There is usually a margin of error in this process. When children have mastered estimating the age from counting the rings, they can try out other ideas: *how big was the tree trunk when the tree was the child's age? How long would it take a tree trunk to grow an inch?* (Remember, the tree grows on both sides of its center.)

Activity 2: Compare Two Tree Cross-Sections

Objective: Children will refine and expand on their ability to count annual rings, interpreting the differences between growth rates on two different trees.

Concept: Tree slices can teach much more than the age of the tree. They can tell the tree's life story.

You Will Need:
- *Contrasting tree slices from at least two different trees.*

One could be from a larger tree, the other smaller; or one from a fast-growing tree and one from a slower-growing one. To assure yourself a good supply of tree slices, tell your original supplier how successful the first lesson was, and how eager the children are to learn more!

What to Do: Have the children compare two different tree slices systematically.
Record this information in a data table:

Width of tree trunk	Estimated age of each tree	Consistency of width of growth rings	Possible reasons for periods of faster or slower growth such as more or less sun, water, etc.

Name

Tree Cookies

A tree cookie is a slice of a tree trunk that foresters and teachers use to show how trees grow. In each cross-section of tree trunk, you can often see rings of varying sizes depending on the growing conditions of the tree.

- **Count the rings on the tree slices below and write the age of each tree in the space provided.**
- **Can you think of some reasons for the differences in the width of the tree rings below?**

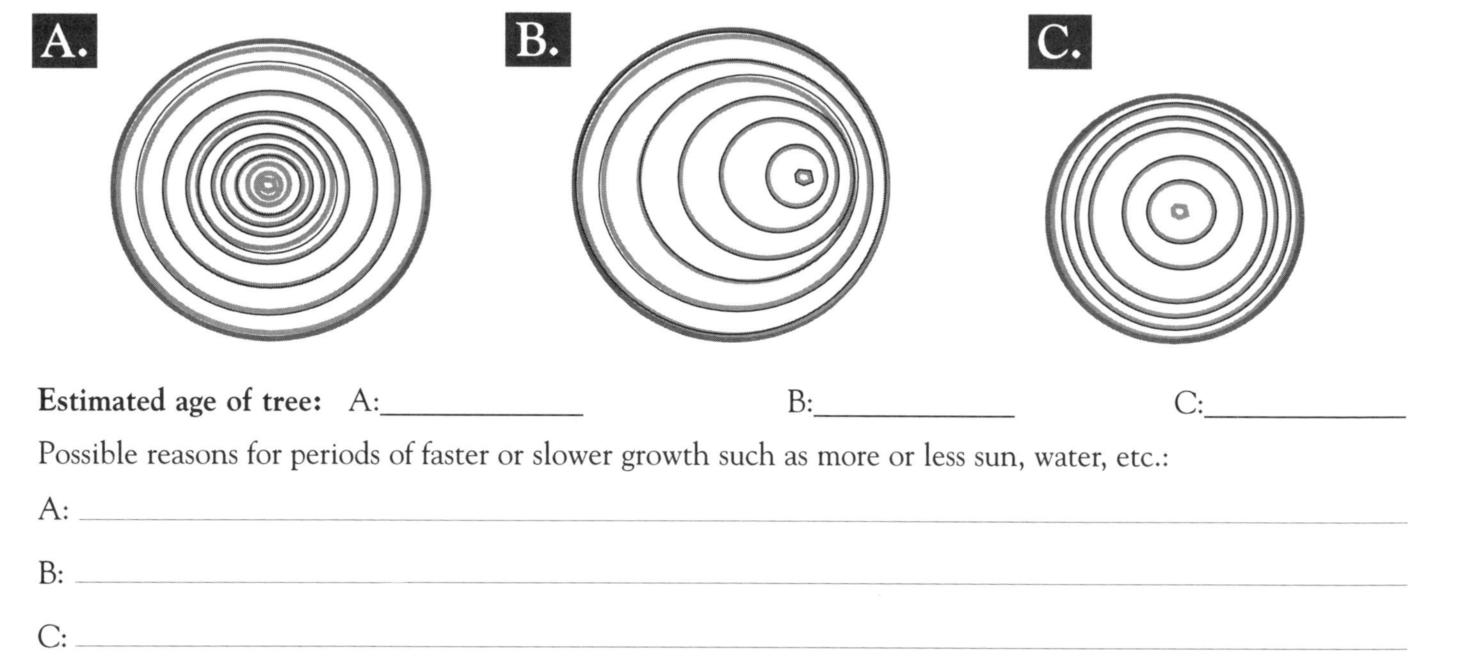

Estimated age of tree: A:____________ B:____________ C:____________

Possible reasons for periods of faster or slower growth such as more or less sun, water, etc.:

A: ______________________________

B: ______________________________

C: ______________________________

Tree Cookies *Here are some you can eat!*

1 c. sugar
1/2 tsp. salt
3/4 c. butter or margarine
1 egg
1 tsp. grated lemon rind
1 tsp. vanilla
3 c. flour
1 tsp. baking powder

Icing
Mix and stir until smooth.
1 lb. confectioner's sugar
1 tsp. vanilla extract
1/3 cup water
green food coloring

1. Preheat oven to 350 F °.
2. Cream sugar, salt and butter. Beat egg, lemon rind and vanilla into creamed mixture until light and fluffy. Sift flour and baking powder and stir into batter.
3. Roll a small amount of dough at a time 1/8" thick on a floured board.
4. Cut with a floured, tree-shaped cookie cutter.
5. Bake on lightly greased cookie sheets for 8 min. or until the edges of the cookies are golden.
6. Cool and frost with icing.

Activity 3: Breathe to a Tree

Objective: Children will learn that a tree makes wood mostly from air and water (and a few other things). They will also learn that a tree gives off oxygen, without which we could not live.

Concept: Trees carry out a chemical change when they make wood. They rearrange the molecules in air and water to make sugar and then wood. Such chemical change is very hard for children (and most adults, as well) to understand. Most children, and adults, too, assume a tree gets what it needs to grow through its roots. But wood is mostly carbon, which comes from carbon dioxide in the air. When wood turns black from burning, you are seeing the carbon. It is counter-intuitive to think that trees make wood from air. It takes a long time to really understand the complex process of photosynthesis, by which plants manufacture food and then wood. Perhaps this experience, and a simplified explanation, can give them a memorable start.

You Will Need:

- *A relatively small tree, with foliage at child height, in a place where children can safely make a circle around it*

What to Do: Indoors, ask children what they know about how trees grow. *How do they make food, and how do they grow?* Most children, and adults, too, think plants get most of their food through their roots. But really, much of wood is carbon, and comes from the carbon dioxide that we (and other animals) breathe. After the discussion, take children outside and circle around the tree, holding hands. Talk about how wonderful it is that this tree can use the carbon from human breath to make food, grow, and make wood. It's also wonderful that we can breathe the oxygen given off by this tree. Then all can enjoy breathing loudly to the tree. Children have a lifetime ahead of them to learn the infinite complexities of the process of photosynthesis.

Activity 4: See a "Snag"

Objective: Children will learn about death, life, recycling in nature, and the value of snags (dead trees).

Concept: The death of a tree provides life for many kinds of plants and animals.

You Will Need:

- *A short, sturdy snag (not a badly rotted one about to fall) in a safe place where children can gather around it*
- *Notebooks*

What to Do: Inside, talk about snags as an important part of the forest. Then, take the students outside to observe the snag carefully and record their findings in their notebooks. What signs of life can they find:
Holes and tunnels of insects? A fungus? A woodpecker hole? An ant or a spider? The claw marks of animals that climbed the tree?

Save a Snag for Wildlife! More than half of all endangered or threatened species in the United States live in forest ecosystems. Snags are needed by different kinds of animals to make their homes.

Name

Crack the Code

Trees are the most important inhabitants of the forest. Even dead trees are important–they provide homes for many kinds of wildlife.

•Try to find out the answer to this Forest Riddle.

Riddle: What do you call three singing trees?

Answer:

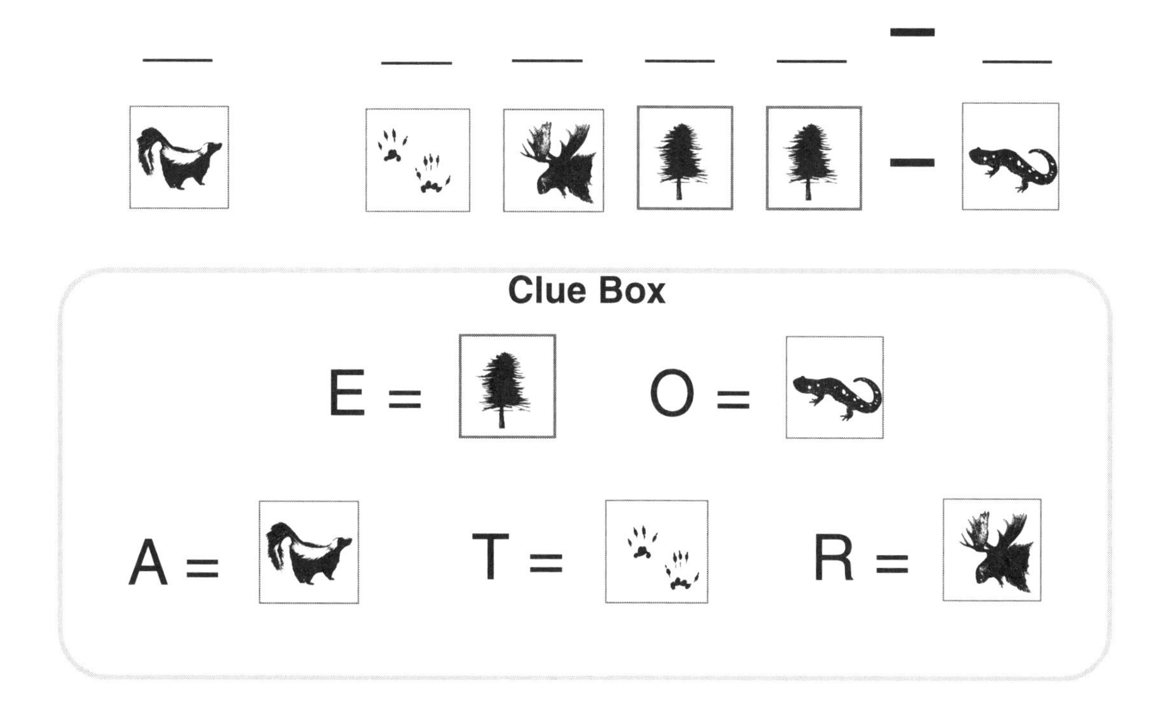

Did you know?

Almost 1/3 of the United States is covered by forests!

Most natural woodland is a mixture of many kinds of trees. Next time you go into a woodland, look up at the trees. Are they hardwoods (with leaves)? Or softwoods (with needles)? Are some very tall? They are canopy trees. Are some trees shorter, living in the shade of the taller trees? These are the understory trees. How many of the trees are dead? How many have fallen down, and are turning into soil?

Name ______________________________

Babes of the Forest

Humans call their offspring *babies*.
There are special names for animal babies, too.

•Use the words from the Word Box to fill in the empty spaces in each sentence.
•Cross off each word as you use it.

Word Box

owlet		
calves	eaglet	bunnies
chicken	tadpoles	antlings
pups	larva	cubs
fawns	kitten	nestlings

1. Young ants are called ___ ___ ___ ___ ___ ___ ___ ___ .
2. Baby bears are called ___ ___ ___ ___ .
3. Baby birds are referred to as ___ ___ ___ ___ ___ ___ ___ ___ ___ .
4 Caribou and elk babies are called ___ ___ ___ ___ ___ ___ .
5. Coyote and wolf babies are called cubs or ___ ___ ___ ___ .
6. Deer babies are usually born two at a time and are commonly known as ___ ___ ___ ___ ___ .
7. A baby eagle is called an ___ ___ ___ ___ ___ ___ .
8. Frog babies are sometimes called polliwogs but usually they are just called ___ ___ ___ ___ ___ ___ ___ ___ .
9. Mosquito young are called ___ ___ ___ ___ ___ .
10. This name is reserved for baby owls. ___ ___ ___ ___ ___ .
11. The name for a baby skunk is the same as the name of a baby cat. ___ ___ ___ ___ ___ ___.
12. Here's one you'll never guess: a baby turtle is called a ___ ___ ___ ___ ___ ___ ___ ; hens have these too and it rhymes with QUICKEN.
13. The rabbit is famous for producing several litters of young a year; 4 or 5 are born at a time and they are called ___ ___ ___ ___ ___ ___ ___ .

CHAPTER EIGHT

Ways of Harvesting

Sound Stewardship of our forests is based on a knowledge of forest ecology and forest succession, knowledge about protection of soil and watersheds, and knowledge of the life cycles and needs of other living things (in addition to trees) in the forest.

Children need to have assimilated the concepts in Chapters 1–7, understand the time frame of tree and forest growth, and be mature enough to understand that people can have different values, before it's worthwhile to try to teach complex and sometimes contentious forest issues. But forest issues are part of our everyday news, and, like Sophie in *Shelterwood*, even young children can be taught the values of stewardship and learn that the forest is a complex community of plants and animals. When properly managed, it can provide wood products, clean water, and wildlife habitat for generations to come.

> *Woodman, Spare that tree !*
> *Touch not a single bough !*
> *In youth it sheltered me,*
> *And I'll protect it now.*
>
> –George Pope Morris

Often economics enter into the choice of which harvest type to use. Clearcuts sometimes cost more to implement (bigger, more harvesting machinery), but produce the highest immediate income. Seed-tree harvests also remove a lot of trees but cost a little more to lay out and administer. Shelterwood harvests require a close examination of the forest stands and require changes in harvest intensity as the forest changes. Selection harvests require single trees to be marked (usually), raising management costs, and only harvest about 25–30 percent of the trees at any one time.

But there are long-range considerations. A quality tree from a shelterwood or selection cut may be very valuable, eventually, as timber. Clearcuts replanted with fast-growing trees and harvested on a short rotation can produce pulpwood for paper quickly, but the wood is not strong enough for saw timber, although some plantations are grown to saw-log size in the South.

When any type of forest harvesting is done carelessly or when the ground is saturated, trees and soil can be damaged. Excessive rutting can be avoided by following Best Management Practices (BMPs). In the North, deep winter frosts fluff up compacted soil, but ruts are unsightly and can contribute to erosion on steep slopes. Siltation of even small streams caused by mud runoff can hurt spawning fish habitat. It can take the forest a long time to recover from these injuries.

Remember the time and care Sophie's grandfather took to protect forest trees and soil. He and Sophie chose "bumper trees" to protect other trees from injury. And Sophie helped lay branches and brush over the ground to protect forest soil from the wheels of the skidder.

Sound stewardship of our forests means understanding their life cycles and protecting their ability to grow and regrow to the benefit of all living things.

Forest Issues

When students are mature and knowledgeable enough about forest ecology to understand complex and contentious issues, they will be interested in learning more about the following:

Large-scale forest harvests are going on now, both in the U. S. and worldwide. Are these harvests sustainable at current levels?

Wilderness areas are large areas where the human influence (roads, motorized vehicles, development) is absent. Such areas are very rare. Are they important? Do we need more of them?

Old growth forest is an uncut forest with very old, large trees. The forest structure is very complex with many layers, and many dead standing and fallen trees. Biological diversity is high, with many more kinds of insects, lichens, fungi, amphibians, and birds than in other forests. Old growth forest is now quite rare. Nearly all of our forests in the U. S. are second growth after being logged. Old growth forest is irreplaceable in any reasonable human time frame. Shouldn't we protect what little remains?

We are currently losing biodiversity worldwide as species become extinct. Shouldn't we leave some places wild enough to support the natural diversity of life? Or will we replace the natural world with a biologically simplified, managed, and developed landscape?

	Shelterwood	Selection
Harvest Type & Description	In a **Shelterwood harvest,** selected trees are harvested in stages over many years. The best trees are left to provide seeds and shelter for new, young trees that naturally regenerate.	Trees are harvested every twelve to fifteen years in a **Selection harvest,** depending on the growing site. Trees of poor quality, as well as mature trees are removed, while future crop trees are selected to grow.

Advantages	Useful for growing high-quality trees used for lumber; regenerates shade-tolerant trees; usually looks nice even right after harvest; no planting costs; enhances the diversity of the ecosystem; provides a diversity of species and tree sizes; creates new habitat for birds, animals, and other plants while leaving needed cover and shade.	Useful for growing the best quality trees for high-value lumber; looks the best right after harvest; creates new habitat openings for some songbirds; provides the best shelter cover for wintering deer while fostering new browse nearby.
Disadvantages	Not a useful way to grow seedlings of intolerant trees; unless the logging is done carefully and at the right time, nearby trees can be damaged during the harvesting operation; deep ruts can damage root systems of the residual trees if the ground is too wet.	Not a useful way to grow seedlings of intolerant trees; most difficult to conduct without causing damage to residual trees or root systems; lowest short-term financial return, highest management costs.
Results	In some shelterwood harvests, all the shelterwood trees are cut down once the new seedlings become well established. In fifteen or twenty years, this will result in a mostly even-aged forest. Some landowners, like the grandfather in *Shelterwood*, use a phased or staggered shelterwood to create a forest landscape populated with a variety of tree sizes and age classes. The grandfather's stewardship really comes closer to managing on a selective basis, rather than a forest stand basis. Both methods are valid, depending upon site and soils.	As selection occurs over several harvest cycles, a variety of age and size classes develop among the growing trees. Done well, the quality of the trees in the forest improves with each entry, usually every twelve to fifteen years.

In a **Clearcut harvest,** all the trees in an area are harvested at once. If young trees are growing under the mature trees, an overstory removal is implemented, leaving the understory thinned out and released to grow.

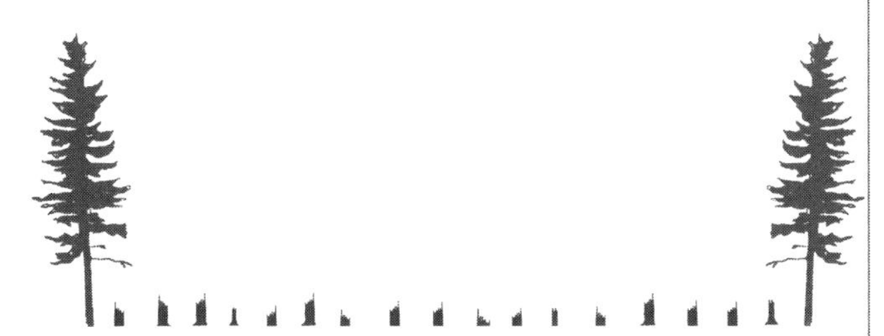

Used for harvesting when an area has mostly diseased or overmature trees; used when the forester wants to create wildlife habitat openings for grouse or hares; used when the landowner wants to replant or regrow a certain species of trees; used by some landowners to maximize short-term profits.

Clearcuts look terrible. Clearcuts raise soil and water temperature, killing amphibians, insects, and other small creatures. They can raise water temperatures in streams, and cause erosion and siltation of fish spawning beds. Clearcuts larger than thirty-five acres disrupt wildlife habitat for some species.

Sun-loving or pioneer species of trees quickly regenerate after a clearcut; these will all be the same age. If planted, usually a single species is grown on the best sites. Poor soils or wet areas are rarely planted. Very large plantations (several hundred acres) lack diversity and are susceptible to being wiped out by insects or disease. Naturally regenerating clearcuts grow through several stages before reaching a climax or late successional stage in as much as one or two hundred years.

Most of the trees are cut in a **Seed-tree harvest,** leaving a well-spaced stand of a few good seed-producing trees. These trees help to reseed the next forest with valuable seedlings.

Provides seeds for new tree growth; provides perch and nest trees for raptors; looks a little better than a clearcut.

If done on a shallow soil, seed trees may be blown over by a windstorm; while widely spaced trees are left, aesthetics are not much better than a clearcut; can cause erosion and siltation if done improperly or at the wrong time.

Later, in ten to fifteen years, the new, young forest develops in the shade of the larger trees; because it is naturally regenerated, it will support a variety of tree species.

Name

Woods Words

These are words that foresters use. Can you figure out what these words are? And do you know what they mean?

•Use the clues to unscramble the words below. Then write the letters of of words in the spaces to the right of the matching clues.

HRAVSET	To gather in a crop	
TNADS	An area of natural woodland where only a single kind of tree is found	
LEACRCTU	Cutting down all the trees in an area	
NOCE	Pine trees store their seeds in these woody structures	
OPT	To cut off the top of a tree	
NOCAYP	The top layer of leaves in a forest	
SIORENO	The removal of soil from the surface by water, wind, or gravity	
DDKSIRE	A tractor-like vehicle used to drag logs out of the forest	
GOLREG	One who earns a living cutting down trees	

Kids can help the forest!

Kids can help the forest by doing these things:

Recycling paper•Buying recycled paper•Picking up trash and litter•Planting a tree in your yard or the schoolyard

Kids can help the forest by NOT doing these things:

Not wasting paper•Not littering•Not carving initials in trees•Not breaking tree branches•Not kicking fungi off tree trunks•Not peeling bark off trees

Activity 1: Doing the Job Well

Objective: Understanding responsible forestry management practices.

Concept: One of the major factors in the long-range success or failure of any particular harvest system is how well it is done. Damaging, careless practices can cause problems and affect forest productivity and animal habitat for years to come.

You Will Need: • *Ample blackboard space*

What to Do: Divide the class into four groups, one for each harvest type. Ask the students to brainstorm "do's" and "don'ts" for each harvest technique, and to list them on the blackboard. Remind the students to think about long-range implications. Older students might incorporate some research into this project and bring their findings back to their group.

"While there may be no "right" way to value a forest or a river, there is a wrong way, which is to give it no value at all. How do we decide the value of a 700-year-old tree? We need only to ask how much it would cost to make a new one, or a new river, or even a new atmosphere."

–Paul Hawken

Activity 2: Forests in the News

Objective: Develop an awareness of forest issues as part of current events.

Concept: Because forests and forest products are very valuable to different people in different ways, forest issues are often in the news.

You Will Need: • *Access to newspapers and periodicals, either with a classroom collection, at the school library, or with materials students bring from home*

What to Do: Divide your classroom into groups of four students and ask each group to find an article in a newspaper or magazine that discusses a forest issue, either in your own community, state, country, or around the world. This research can be done over a period of a week or two. Older children can read the articles themselves, but younger children may need to work briefly with a teacher or parent to understand the material and be able to summarize it. Ask each group to prepare a short news broadcast about the news item, with the "anchor team" reporting the basic facts, and then "specialists" analyzing it further, relating the facts of the news item to what they've learned about habitat, tree growth, and harvesting techniques.

Name

AMazing Forests

When forests are carefully logged, more trees are saved for the future. Careless logging can cause severe soil erosion and disrupt wildlife habitat.

- **Enter the maze at the top left and find your way through the forest. Beware of forest practices that might harm the forest or the environment.**

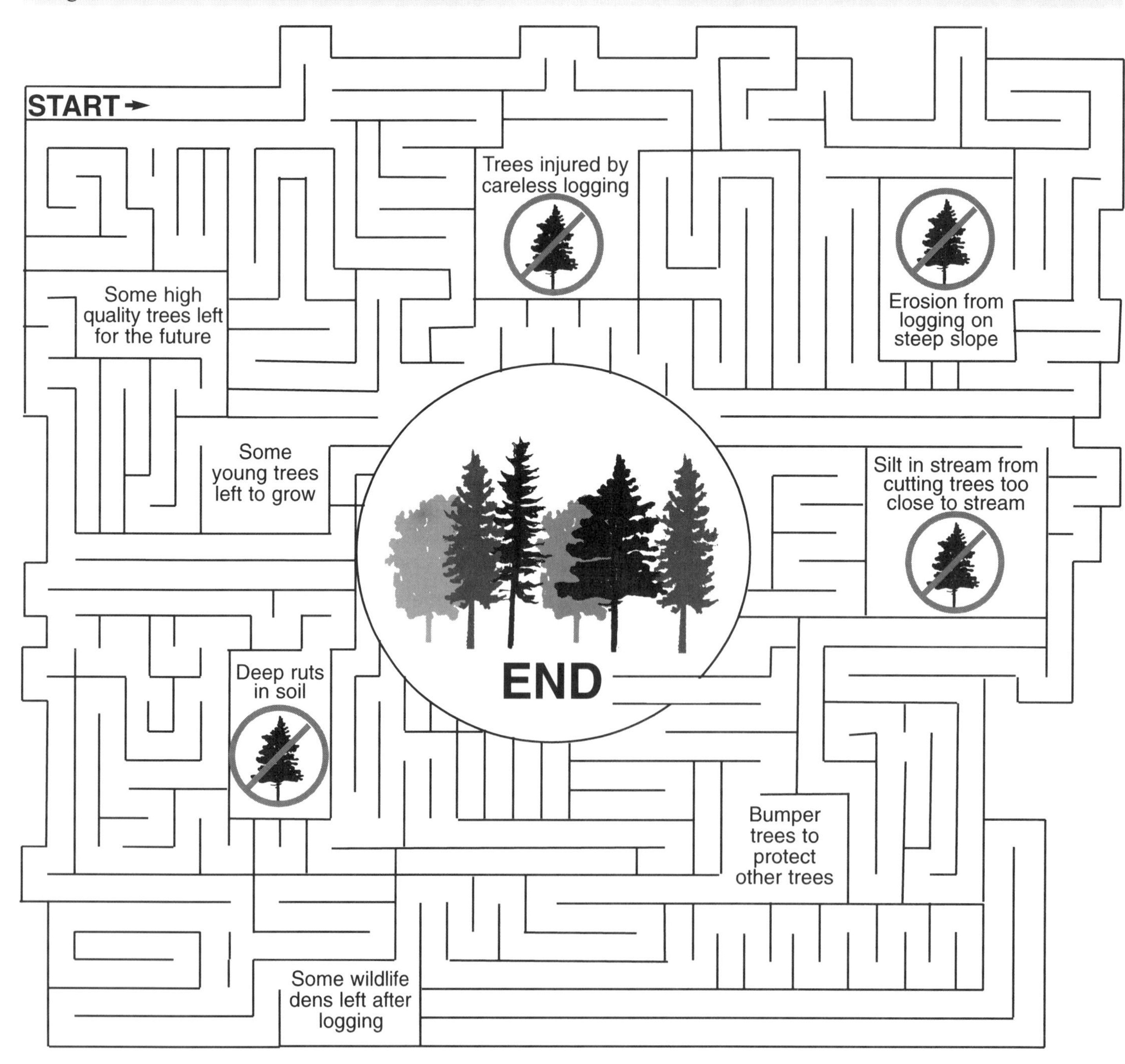

CHAPTER NINE

Boards, Furniture, Paper, and Even Violins

People use trees for many things—boards, paper, furniture, violins, and much more. Don't forget that every minute of our lives, we are breathing the oxygen from trees—without that, we couldn't survive.

Look around your room carefully. How many things can you find that are made from trees? Things made from unpainted wood are immediately obvious. It is fun and educational to look carefully at all these objects. Can you find evidence of growth rings throughout the wood? Is your desk solid wood, or veneer, or a combination of veneer and composite? (Veneer is a thin layer of wood, often overlaid on a composite of resin and tiny wood particles.)

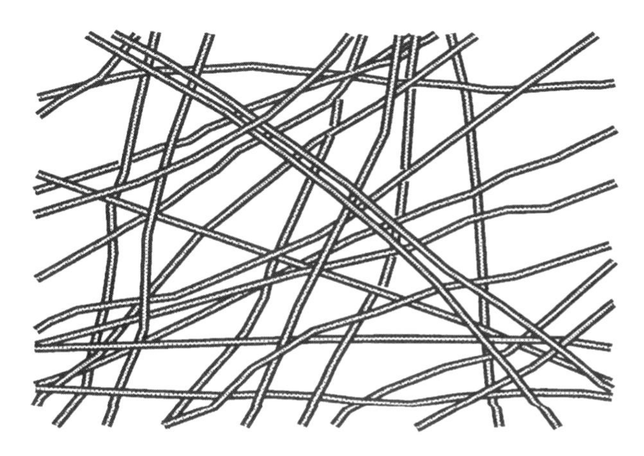

Most paper is made from the long, strong fibers of softwood trees.

A knowledgeable person in a furniture store can help you learn to differentiate solid wood from veneer and composite. Children will be fascinated by these differences. A friendly lumberyard salesperson might even give you small samples of solid wood, veneer, plywood, and several kinds of composite, perhaps showing the differences between fine and coarse wood particles. Not many teachers teach this, but children are invariably fascinated because it is concrete and hands-on.

A music teacher might be willing to bring a violin into class, showing the different kinds of wood that are used in the different parts of a violin, and even the rosin for the bow. Can children hear a difference in the sound of a pre-rosined and post-rosined bow?

It is not intuitively obvious that paper is made from wood. It helps to explain to children that paper is made from wood fibers, which (in wood) are like stiff, straight threads. A strong magnifying glass can help children see the fibers in a broken edge of a piece of wood. (They should be careful of splinters! How many children have had the painful experience of a splinter? A splinter is made of wood fibers!) The wood fibers in paper can best be seen on a torn edge of paper with a strong magnifying glass. Try this also with cereal box cardboard and corrugated cardboard.

Hardwood trees make good tissue paper and paper towels. The fibers of hardwood trees are short and soft. Softwood trees make good, strong, book and magazine paper. The fibers of softwood trees are long and strong.

The flattened fibers can easily be seen in this greatly enlarged depiction of a paper's surface.

Linoleum flooring, once eclipsed by vinyl but now enjoying a resurgence in popularity due to its natural origins, is made from pine resin, wood flour, and other natural substances. Perhaps a friendly flooring salesperson can give you small samples of vinyl and linoleum flooring to let children compare. They can feel and sniff both samples, and look carefully at the cut edges with a magnifying glass. It is not intuitively obvious that linoleum comes from trees, but it does!

Children may also forget that many fruits and nuts come from trees, as do maple syrup, resin, rosin (for the violin bow), and turpentine. What else can they think of that comes from a tree? How long a list can they make?

Activity 1: How Much Can You Learn from a Board, without Getting Bored?

Objective: Children will learn how a board is sawed from a log, what makes a knot or a knothole, how the growth rings go all the way up the tree, and more.

Concept: Children will enjoy looking at a board, figuring out how a board would be cut from a log, studying why the growth rings look different on a board from a tree slice (see Chapter 7). On the ends of the board, you can see some of the growth rings and imagine how this board was cut out of a log.

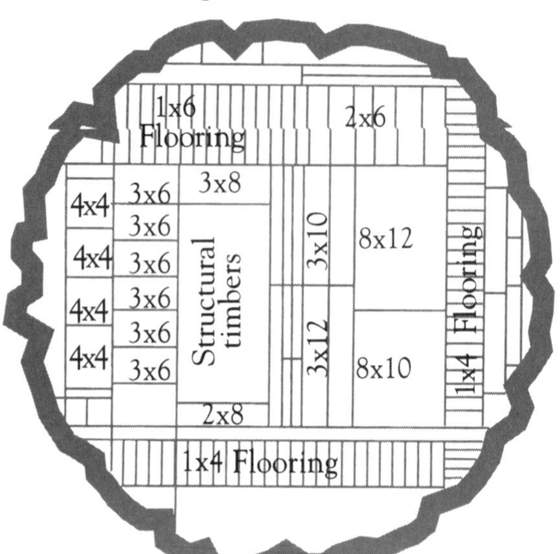

Crosscut, showing how boards are cut from the log.
The largest sections are sent to the sawmill. Smaller and lower quality pieces are chipped and sent to paper mills.

You Will Need:
- *Short, smooth pine boards, without splinters, a foot long or less, some with knotholes*
- *Paper and pencil, for the children to report their findings*

Two or more children can share each board. They must be cautioned about careful handling of a board so that no one gets knocked by a sharp corner. A lumberyard that does sawing, or a construction worker or carpenter can likely supply you with some good, small board scraps for educational use.

What to Do: Pass out the boards.

Where does a board come from? What is it made of? Can children find the growth rings? Can they draw a tree slice (see Chapter 7) *of what the original log must have looked like, with a rectangle of where this board would have come from inside the log?*

How many years of the tree's life are in this board?

Does anybody have a board with a knot in it? Do they know what a knothole is? Let them guess. If they have trouble guessing, a knot is a branch that grew when the tree was young. Later, as the trunk grew thicker, it grew around the branch. The branch was then hidden inside the tree trunk. When the trunk was sawed into boards, the branch became visible again in the board, as a knot. The knot has its own growth rings, formed as the branch grew. Sometimes the knot falls out of the board, and then you have a knothole. A knot weakens a board; the strongest boards, posts, and beams have no knots.

Name

Products We Get from Trees

Lots more than wood and paper come from trees!
More than 5,000 commonly used products come from trees.

- **Hidden in the forest puzzle are just a few products that come from trees. See if you can find and circle them in the puzzle below. The words may be spelled up and down, sideways, backward, or diagonally.**

Word Box

paper	fruit	soaps
furniture	crayons	shingles
charcoal	chewing gum	baskets
syrup	perfumes	tea
nuts	medicines	lumber

C	W	P	L	E	S	T	E	K	S	A	B
O	H	H	E	A	D	P	A	P	E	R	P
L	S	E	S	R	O	S	E	S	L	Q	U
Z	C	E	W	R	F	C	T	F	G	V	R
F	U	R	N	I	T	U	R	E	N	F	Y
N	X	E	A	I	N	H	M	A	I	O	S
J	R	B	E	Y	C	G	T	E	H	O	K
E	B	M	J	W	O	I	G	O	S	C	E
T	I	U	R	F	X	N	D	U	W	E	I
D	G	L	S	P	A	O	S	E	M	Q	M
O	J	S	T	Q	N	N	P	M	M	R	Z

Why do we need to use recycled products whenever we can?

It takes a long time to grow trees that are large enough for making paper.
Most paper is made from 30- to 80-year-old softwoods like pine trees. Hardwoods–like oak and maple trees–can take over 100 years to grow.

Activity 2: Making Recycled Paper

Objective: Children will learn how paper is recycled.

Concept: Paper is made from tree fibers. Once used, paper can be recycled into new paper, saving trees. Children enjoy recycling newsprint into a different kind of paper in this hands-on process. In big factories, a similar process is used but on a much larger scale. Teach children also that they should recycle paper at home and in school, and when they buy paper, they should look for recycled paper, especially paper with "post-consumer" recycled content. Explain the difference between "recycled" and "post-consumer recycled" paper to your students.

You Will Need:

- *An old newspaper*
- *Cheesecloth*
- *Shallow tin cans (like tuna fish or cat food cans) with both ends removed*
- *Rubber bands*
- *Absorbent dishcloths or dishtowels*
- *Several sponges*
- *Rolling pin*
- *Blender**
- *A quart of water*
- *Food coloring (optional)*

What to Do:

1. Cut four pages of newspaper into 2-inch squares.
2. Put squares in a bowl and cover with warm water; soak for an hour.
3. Then, put soggy paper a little at a time, with water, into the blender; blend to form a thin, frothy liquid—"paper soup." If you want, you can add food coloring to make tinted color. If you add no color, the paper will come out gray.
4. Cut the cheesecloth into 6-inch squares. Stretch several squares of cheesecloth across the bottom of each tin can, attaching it with a rubber band.
5. Put one-half inch of water into a shallow pan.
6. Put the cans in the pan, cheesecloth side down.
7. Pour a cup of the paper soup into each can. Swish it gently, lift the can out of the pan, and let the water drain into the pan.
8. Place the can on an absorbent dishcloth or towel.
9. Take the rubber band off and take the tin can off. *You will have a flat "paper cookie."*
10. Press it with a sponge to absorb as much water as you can; your new paper "cookie" should come off the cheesecloth. Put the paper "cookie" between two dishtowels and roll with a rolling pin to flatten it. Let it dry overnight or, preferably, over a weekend and you'll have a piece of newly recycled paper.

When admiring the final product, again remind children about the importance of recycling paper at home and school, and of buying paper with post-consumer recycled content.

***Note: Be aware that this activity can be hard on the blender.**

CHAPTER TEN

Seeing Wildlife

Most children love animals and are fascinated by them. Their first experiences with animals are usually with pets, or domestic animals. These can be seen close up, patted, played with. In other words, children can have close-up, hands-on, and emotionally involving experiences with pets.

Children must take a conceptual and emotional leap in learning how to relate to wild animals. It is inappropriate, and sometimes dangerous, to try to play with them or pat them as one does with a pet.

> *That is the charm of the woods, anyway.*
> *Things live and breathe quietly and out of sight.*
>
> *-Lucy M. Boston*

Children must learn to generalize from the love of an individual pet, to a more abstract caring about the forest and its inhabitants. This generalized love of the forest is exemplified in the book *Shelterwood* by the grandfather as he teaches it to Sophie in many ways.

Sophie has a pet cat, but she leaves her cat behind as she goes into the woods with her grandfather. She is thrilled and satisfied to get quick glimpses of wild animals. She has learned to love their wildness in itself, and to be satisfied with a quick glimpse or with finding an animal's footprint, the secret sign of its presence.

Sometimes when children go into a forest setting, their hopes are unrealistically high about seeing moose, deer, or bear. There are many animals in the forest, yet one can easily go for a walk in the forest without seeing one.

There are two important forest lessons for children here:

- The forest is a good habitat for animals because it has so many shelters and hiding places that make them hard to see. Also, many animals sleep during the day and come out only at night.

- Children must therefore be detectives and know how to read and interpret the secret signs of animals.

Activity 1: Hiding Places in the Forest

Objective: Children will learn that the complex structure of the forest offers many homes for other living things.

Concept: The complexity of forest structure—trees, live and dead, standing and fallen; logs on the ground; holes in trees, among rocks, and in soil; dense shrubs and thickets—creates excellent shelters and hiding places for forest wildlife.

You Will Need: • *Big rolls or sheets of paper for a wall mural*

What to Do: Have children make a forest mural depicting and labeling all the elements above. Then they can research common local forest animals, draw and cut out pictures, and attach them to the mural. They can create ingenious ways of hiding their cutouts behind a flap of tree bark, under a rock, inside a thicket, etc. as they learn about the animal's habits. Be sure they include not just mammals, but also birds, reptiles, and amphibians like tree frogs, if they occur in your area. Insects and even stream fish can be important parts of the forest wildlife community, as well. Children might like to include people in their mural as part of the forest, using the forest for recreation or for forest products. People are part of forest ecology, too. And, if they put themselves in the mural, children can be pleased that they are part of the forest.

Now, on a visit to the forest, children will enjoy finding some animal's secret home instead of being disappointed about not seeing the animal itself.

Note: Safety information is important—children may want to reach impetuously into holes in the trees or among rocks, but they should be advised ahead of time against this.

Activity 2: Hiding Places in a Tree

Objective: Children will learn that a large tree itself is complex in structure, offering many homes for other living things.

Concept: Any large tree is full of hiding places. Cracks in the bark can hide insects. Crevices in the trunk can hide a flat, slim animal. A large hole sometimes forms where a branch broke off many years ago. An old woodpecker hole can now shelter a mouse, as can an old bird nest. Squirrels build nests of sticks in trees, just as larger birds do. Twigs, leaves, and roots can hide many kinds of animals.

You Will Need: • *A clipboard or notebook*

What to Do: Have children go to any large tree, or the tree they selected for the activity "Keep a Tree's Diary" in Chapter 2. They will use their growing knowledge of local wild animals and their respective sizes. They can look for places these animals theoretically could hide on or in this tree. Young children may not realize that an animal's size determines where it could hide: bigger animals need bigger trees!

Name

Points About Porcupines

Porcupines are good-natured creatures and are not aggressive. They prefer not to fight, and will attempt to escape up a tree.

•Here are some interesting facts about the porcupine:

- A porcupine has 30,000 quills.
- A porcupine turns its back to attack and *covers its face* with its front feet.
- *New* quills soon grow in to replace *used* ones.
- The porcupine is our *second largest* native rodent. (The beaver is largest.)
- A porcupine's claws are strong and *curved*, for climbing trees.
- Porcupines are *good swimmers* because their hollow quills help keep them afloat.
- Baby porcupine quills are *soft* at birth; but they harden in a few minutes.
- Baby porcupines are called "*porcupettes.*"
- A baby porcupine can climb trees when less than *1 hour* old.
- A porcupine's *home* is in a hole in the ground or in a hollow tree.
- Although it appears slow, a porcupine strikes with *lightning speed*, using its clublike tail.
- Porcupines are *vegetarian*. One of its favorite summer foods is clover.
- Porcupines *don't hibernate* in the winter.
- The porcupine curls up in a furry ball near the *top of a tree* to sleep in the daytime.

Name

Skunk Sense

Below are some interesting facts about skunks.

•Circle the facts that are new to you.

More skunks perish on the highways than from any other cause.

Skunk musk is luminous when sprayed in the dark.

Skunks have long nails especially on the front feet, for digging.

Fighting skunks don't seem to use musk on each other.

Skunks are shy and unaggressive animals.

The young often follow behind their mother in single file in their daily trips for food.

The striped skunk is active year-round except in extremely cold weather.

A baby skunk is called a kitten.

A skunk can spray up to 15 feet for 5 to 6 consecutive rounds.

The skunk is currently one of the chief carriers of rabies in the United States.

Skunk musk has been used commercially as a base for perfumes because of its lingering qualities.

In winter, skunks will use abandoned dens or even share a burrow with other animals.

Skunks are boldly colored to advertise to enemies that they are not to be bothered.

Skunks are resistant to snake venom, and it takes ten times as much to harm it as another animal the same size.

There was a young man from the city,
Who saw what he thought was a kitty.
Saying, "Nice little cat,"
He gave it a pat . . .
They buried his clothes out of pity.

CHAPTER ELEVEN

Reading the Secret Stories Told by Animal "Signs"

Animals stay hidden so easily in the complex structure of the forest that anyone who wants to know their habits must learn to read their signs.

It takes a long time to become proficient in wildlife tracking, but children love its challenge. They will plunge right into it, gaining skills and confidence as they go along. You can tell them that they will keep learning about the forest all their lives, because nature reveals its secrets only slowly. Tracking is best learned by an ongoing combination of reading and experience.

There are many wonderful books for children and adults alike about animal tracks. When you go looking for tracks, it helps to have a working knowledge of what animals are most common in your area. Many natural resources organizations have information about the most common forest animals and their tracks.

Then, go out and look! Tracks can be found in the muddy edge of a pond, in the dry sand, or in snow. The challenge with the field study of tracks is that children get very excited about them and the tracks themselves often get trampled in the excitement. It sometimes helps to find a good tracking area ahead of time and put in a line of stakes, beyond which children should only look, not walk.

Children can be asked to look for tracks in their own neighborhood, make a life-sized drawing of the tracks, and bring the drawing to school to show and tell about.

And tracks are not the only "signs." Others are:

• A nest or home, such as a bird nest or a beaver lodge, or the hole of a squirrel;

• A shed hair or feather, or a cast snakeskin;

• Scat (a scientific term for feces). Most tracking guides also help identify scat. Children should not touch it, of course, but they will be fascinated to learn that scat can be identified, and that scat often indicates what the animal ate;

• Signs of feeding, such as an acorn shell chewed apart by a squirrel;

• Claw marks on smooth trees. Such claw marks are often left by squirrels or raccoons.

Activity 1: Be a "Track Detective" in the Classroom

Objective: To learn to observe human tracks, footprints, and patterns in an indoor setting.

Concept: Children's feet and shoes differ, just as feet of different animals do.

You Will Need:
- *Aluminum foil*
- *A thick towel*

What to Do: Select three children whose shoes all differ. Tell them to come with you just outside the classroom door, or someplace in the classroom, where other children can't see them. Select one child who will step on a 1 x 1-foot piece of aluminum foil placed on top of a folded, thick towel. This should make an imprint in the aluminum foil. Take all three children back in view of the class. Have them turn their backs to the class and hold one foot up so the sole is in view. Hold up the aluminum foil imprint. *Can children guess which child made the footprint?* (You can adjust the difficulty level by choosing children with more or less similar footwear.)

Activity 2: Draw a Story Told by Tracks

Objective: To learn how to make and interpret a track "story."

Concept: Tracks really tell a story, one which takes some creativity to interpret. Most animal track stories are simple, without much plot.

Here are typical examples:

A mouse hops out of a hole, goes to a berry bush, and then back to its hole

A squirrel climbs down a tree, hops across the snow, and climbs up another tree

A rabbit hops from under one dense bush to another

You Will Need:
- *Paper*
- *Art materials*
- *Guidebooks illustrating different kinds of tracks and track patterns*

What to Do: Make a simple drawing of several simple track stories. Show them to the children, explaining how you invented the story and figured out how to draw the tracks. See whether children can figure out several of their own track stories. Each child can then show and tell his or her stories to the others. With practice, the stories can get more dramatic or fanciful.

Name

Tricky Tracks

Who lives in the forest? By looking closely at animal tracks, you can often tell where they were going and what they were doing. If the track has claw marks, they will point you in the direction the animal was going. Mud and soft soil are great places to find animal tracks.

- **Look at each set of tracks.**
- **Guess who made each set of tracks by checking the box beside one of the three choices.**

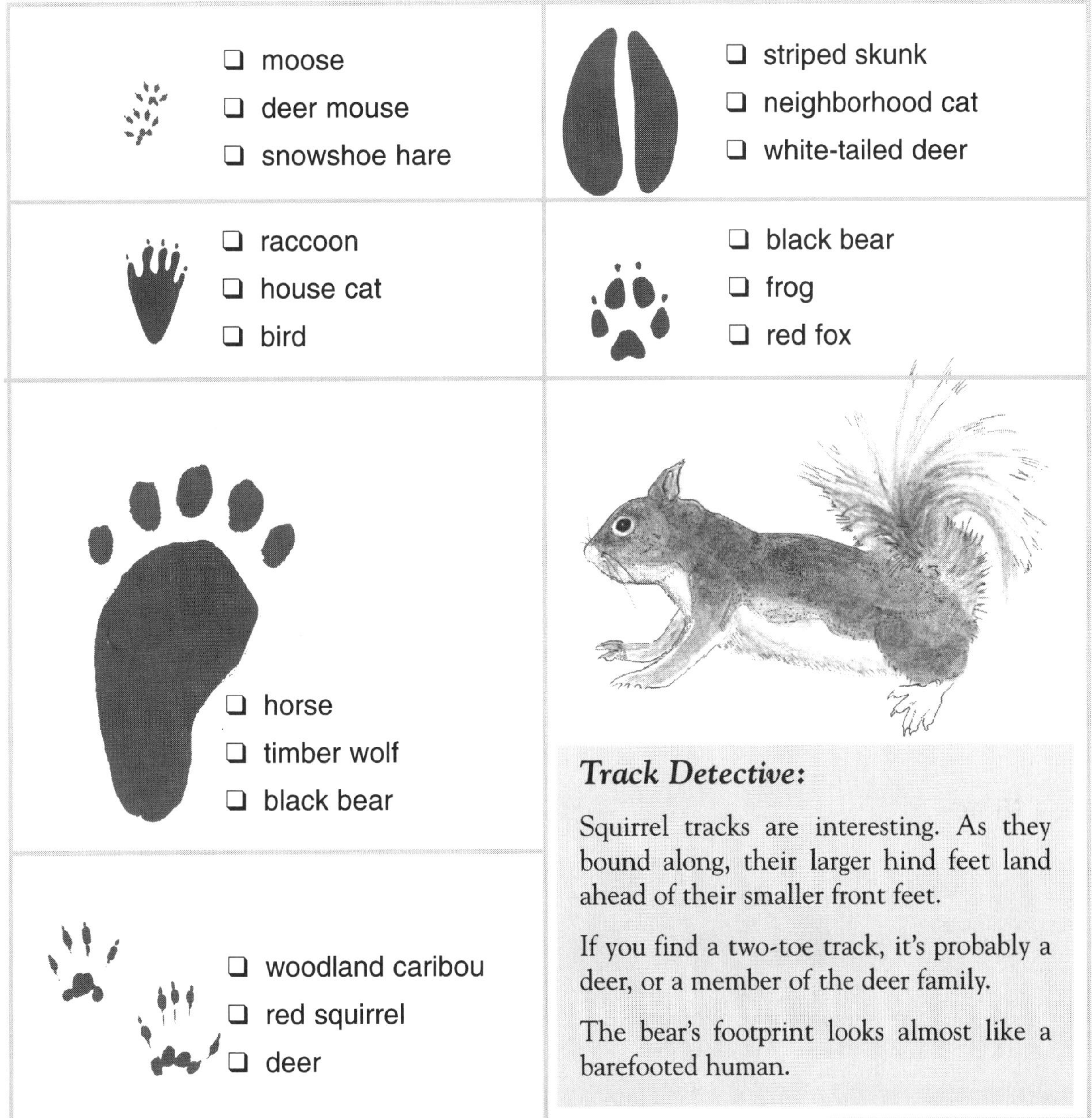

Track Detective:

Squirrel tracks are interesting. As they bound along, their larger hind feet land ahead of their smaller front feet.

If you find a two-toe track, it's probably a deer, or a member of the deer family.

The bear's footprint looks almost like a barefooted human.

Name

Life in the Leaves

The thick layer of dead leaves that settles onto the forest floor each year helps protect roots from freezing temperatures. It also provides homes for many types of small animals.

•Unscramble the words below to discover a few of the animals of the forest floor!

c m e i — — — —

r e d n m a s l a a — — — — — — — — — —

t s c e i n s — — — — — — —

m e r a h t o w r — — — — — — — — —

e l o v — — — —

n k s a e — — — — —

Chapter Twelve
Owls

The little screech owl's voice is often heard, but many people don't realize it's an owl because it doesn't hoot or screech. Instead, it gives a little whinny, then a trill.

Children love learning about owls! An owl's eyes both face front, just like the eyes of people. (The eyes of most birds are on the sides of their head.) An owl has "binocular" vision, meaning that it focuses both eyes on the same object, unlike many birds. Binocular vision helps the owl locate its prey with precision, even in the dark.

An owl's ears are fascinating, too. They are just openings in the sides of the owl's head. An owl's ears are not symmetrical—one is higher, one is lower, and the shapes are different. It is thought this helps the owl hear sounds a little differently in each ear, enabling it to locate prey with precision, even in the dark. (Ear tufts are just feathers; there are no ears inside. It is likely that the ear tufts help break up an owl's outline and help it blend in with a nearby tree trunk.)

An owl's feathers are fascinating, too. They are unlike the feathers of any other bird. Each feather has a velvety layer that looks like fur, and the soft outer edge of each feather looks fringed. These adaptations enable the owl to fly in silence, its approach unheard by prey.

An owl catches its prey (mostly mice, but also rabbits, flying squirrels, large insects, and small birds) with its strong feet and long, curved, very sharp claws which can go right through a mouse like a sword. Before it catches its prey, it has to locate it precisely with its eyes (binocular vision) and asymmetrical ears. Then, with its soft, velvety feathers, it flies silently towards its prey and grabs it.

Activity 1: What Is an "Owl Prowl"?

Objective: To acquaint children with the mysterious, exciting, and fun concept of an "owl prowl."

Concept: While learning about owls, students love the idea of going on an "owl prowl," a nighttime walk looking and listening for owls. It is challenging to take children on a real owl prowl, since it occurs at night, outside of school hours. Owls are generally uncommon and hard to find. There are also safety issues to consider when taking children outside at night.

A Nighttime Owl Prowl

is led by someone knowledgeable about owls. Everybody must be very quiet, a challenge with any group. The leader may play a tape of an owl's hoots, or imitate an owl by voice, in hopes the owl will hoot back or fly in close enough to see. It is somewhat disturbing to the owl, which thinks another owl is intruding on its territory. This may be acceptable where there are few people and much owl habitat, but not where there are many people who may each try to imitate and see an owl in the same area. Disturbing an owl more than once amounts to owl harassment. On the other hand, children (and adults, too) are thrilled to see or hear a real owl. It is up to you to balance responsibly the educational benefits of having an owl prowl with the potential of disturbing an owl.

These are instructions for having a "Simulated Owl Prowl" during daytime hours:

You Will Need:

- *First, read* Shelterwood, *and discuss how Sophie is at first frightened when she hears the owl's call, and then reassured by her grandfather.*
  Another book that relates well to this experience is the beautifully illustrated, award-winning picture book Owl Moon *by Jane Yolen (see references).*
- *You will need a tape or CD of a great horned owl's hoots (see references)*
- *A tape or CD player*
- *A safe room that can be darkened*
- *A light that you can aim and focus on the book, as you read and show it to the children.*

What to Do: Have the children sit down and be very quiet, darken the room, tell the children that you will go on a "simulated owl prowl," and read them the story.
Afterwards, play a tape or CD of a great horned owl's voice.
Can children pay careful attention to the owl's rhythm and learn to imitate its hoots?

Name

Amazing Owls

Owls are fascinating inhabitants of the forest.

•Circle the facts about owls that are new to you.

After they eat, owls cough up pellets made of the fur and bones of their prey.

An owl's talons sometimes go right through a mouse, like a sword.

Sometimes crows gather around an owl and "caw" at it incessantly. This lets all the other forest creatures know where the owl is! This behavior of crows is called "mobbing."

An owl's eyes are very big for its size. This helps the owl see well in the dark. But they can see in the daytime, too.

A larger kind of owl (like a great horned owl) will sometimes kill and eat a smaller kind of owl (like a screech owl).

Short-eared owl. The ear "tufts" are just feathers. An owl's ears are actually hidden openings in the sides of the owl's head.

Most owls do not make their own nests, but use a hole in a tree or another bird's nest.

Owls can't roll their eyes from side to side like people do; instead, they turn their head when they want to see to the side.

The feathers on an owl's face funnel sounds into its ear openings. The facial feathers create a "facial disc" which is like a parabolic reflector, or a satellite disc.

The female owl is bigger and stronger than the male.

Owl's feathers are very soft and velvety. This enables them to fly silently. When most birds fly, you can hear the "flap, flap" or "whoosh, whoosh" of their wings. But not the owl! It flies silently; the mouse never hears it coming.

You can hear an owl call from more than a mile away.

Activity 1: What Is an Animal Family?

Objective: Children will learn about the scientific concept of "family."

Concept: In biology, an animal family is a group of related species which have many features in common and probably evolved from a common ancestor. Children may at first be confused; this is a more abstract definition of family than the one they are used to.

You Will Need: •*Pictures of any of these: great horned owl, barred owl, spotted owl, screech owl, other owls.*

What to Do: Show children pictures, starting with two species at a time, and ask, "*What are the differences? What are the similarities?*" Then line up all the pictures you have and ask the same thing. Conclude by talking about each species as a member of the owl family.

Whooo gives a Hoot?

A **great horned owl's** hoots sound like this: ***hoo, hoo-hoo-hoo, HOO HOO.***

The **barred owl** sounds like this: ***hoo, hoo, hoo, hoo; hoo, hoo, hoo–hooo–aw!*** *(To the rhythm of "Who cooks for you, who cooks for you–all")*

The **screech owl** doesn't screech; its voice is a tremulous, descending whinny, followed by a soft trill.

Activity 1: Which Member of the Owl Family Lives in Your State?

Objective: Children will learn to use the concept of "range."

Concept: The range is the broad area over which a wild animal is found. Young children (and many adults, as well) do not realize that wild animals have a range and are not normally found outside of that range. Any field guide to birds, mammals, or other groups includes range maps of each species. In learning how to identify a species, it is very helpful to know which ones inhabit your area. It is sometimes difficult for scientists to know where the current "edge" of a species' range is. Ranges sometimes change over time. Within an animal's range, it will only live where habitat is suitable. Also, some owls migrate, and may be in your area only in a certain season.

You Will Need: •*Children will need map-reading skills; they will have to recognize a map of the U. S. and know how to find their own state on it*

•*A field guide to birds that includes range maps for owls*

What to Do: Show children the range of different members of the owl family. *Which ones live in your state? Is your state at the edge of any species' range? Have your students ever seen an owl?*

Chapter Thirteen
Deer and Moose

It's a treat to see graceful, wary deer. They are fascinating animals to study. Students from hunting families may have a good start in learning about deer, and all students will enjoy learning about them in more depth.

Deer and moose are in the same mammal family,

The North American moose is well-adapted to deep snow and and a cold climate, and is a symbol of the northern coniferous forest.

the deer family. Members of this family also include elk and caribou. All have cloven hooves and NO TEETH in the top jaw. All males grow antlers every year, and shed them every year. All are plant eaters. They eat by grasping a twig or plant between the upper "gum pad" and the lower teeth. Then they break off the plant against their bottom teeth with a jerk of the head.

Deer live in most of the U. S., while moose live in most of Canada. There are two kinds of deer—mule deer, common in the West, and white-tailed deer, more common in the East. Elk live mostly in the West, while caribou live mostly in northern Canada.

White-tailed deer are perhaps the most familiar members of the deer family. They browse on shrubs and low trees along wood and road edges, and sometimes are easily seen. You may have seen them bound away, white tails waving in the air. Their tails are actually a signaling device, meaning "Danger! Follow me!" to other deer.

If you see a deer cross the road, slow down–often another deer is following right behind! Many deer are hit by cars.

If a deer is startled, it will often make a loud snorting noise as it bounds away.

Did you know that white-tailed deer living in the northern U.S. or Canada are typically larger than Southern deer? (A large body holds in more heat and helps survival in a northern climate.) Tiny white-tailed deer called "Key deer" live on the Florida Keys.

Deer grow a thick coat of long hair in winter. Each hair is hollow. The air spaces inside each hair

White-tailed deer are perhaps the most familiar members of the deer family.

are good insulation and help the deer keep warm in winter.

Moose are less familiar; in *Shelterwood,* Sophie did not see one but found its large, long footprint. If she is patient and spends much time in the Maine woods, she will certainly see one. Their populations are expanding. Moose are a most improbable-looking animal—dark, leggy, and enormous, with a long and bulging nose. They are a symbol of the northern coniferous forest, well-adapted to deep snow and a cold climate.

Activity 1: What Is an Animal Family?

Objective: Children will learn about the scientific concept of "family."

Concept: In biology, an animal family is a group of related species which have many features in common and probably evolved from a common ancestor. Children may at first be confused; this is a more abstract definition of family than the one they are used to.

You Will Need: • *Pictures of any of these: mule deer, white-tailed deer, moose, elk and/or caribou.*

What to Do: Show children pictures, starting with two species at a time, and ask, "*What are the differences? What are the similarities?*" Then line up all the pictures you have and ask the same thing. Conclude by talking about each species as a member of the deer family.

Members of the deer family use their back teeth to twist off twigs to eat. Notice the absence of teeth in the front portion of this deer's upper jaw.

Activity 2: Which Member of the Deer Family Lives in Your State?

Objective: Children will learn to use the concept of "range."

Concept: The range is the broad area over which a wild animal is found. Young children (and many adults, as well) do not realize that wild animals have a range and are not normally found outside of that range. But any field guide to birds, mammals, or other groups includes range maps of each species. In learning how to identify a species, it is very helpful to know which ones inhabit your area. It is sometimes difficult for scientists to know where the current "edge" of a species' range is. Ranges sometimes change over time. Within an animal's range, it will only live where habitat is suitable.

You Will Need: • *Children will need map-reading skills; they will have to recognize a map of the U. S. and know how to find their own state on it*
• *A field guide for mammals that shows range maps*

What to Do: Show children the range of different members of the deer family. *Which ones live in your state? Have students seen them? Is your state at the edge of any species' range? Does the range map concur with student observations?*

Name

All Kinds of Deer

An animal family is a group of related species which have many features in common and probably evolved from a common ancestor. The deer family includes the mule deer, the white-tailed deer, the moose, the elk, and the caribou.

- **Here are some interesting facts about deer. Circle the facts that are new to you.**

The white-tailed deer uses its tail to communicate with other deer: a tail up means "run for cover."

Deer have ears that can twist back and forth to hear even very faint sounds.

A deer's nose is 100 times more sensitive detecting smells than a human nose.

Mule deer can have long ears–like a mule! Their ears can be up to a foot in length.

Mule deer run with all their feet leaving the ground at once, as if on a pogo stick. This is called stotting.

Because their big eyes are on the sides of their head, deer can see what is happening ahead of them or behind them, and most of what is going on behind them, too!

Reindeer and caribou are the same animal (*Rangifer tarandus*). Reindeer is the European term and caribou is the American term.

The mule deer can move each of its ears in different directions at the same time.

A deer has such good hearing, that it can detect how far away an object is just by hearing a sound from that object.

Both male and female caribou have antlers, but the cow's antlers are much smaller. Only the males of other deer have antlers.

Biologists can tell how old a deer is by looking at its teeth. They count how many "baby" teeth are left, or how much the "adult" teeth, or molars have been worn down. They sometimes even look at a cross-section of the deer's tooth under a microscope and count the annual growth rings!

The glands of a deer give off a scent which help it to communicate with other deer.

Name

Moose Math

The moose is found in the forests of North America, northern Europe, and Asia. During the summer, moose can often be found standing in deep pond or lake water munching on water lilies and pondweeds.

- **Work the math problems to find the numbers that are missing from the sentences.**
- **Write the correct numbers in the blank spaces to learn some interesting facts about moose.**

1.The moose is the largest member of the deer family, and may weigh over (500 x 3) - 100 _____ pounds!

2. Moose can run up to (6 x 5) + 5 _____ miles an hour.

3. The tracks of the bull, or male moose, can be (4 x 3) - 5 _____ inches long.

4. At five months old, the moose calf may stand (6 x 12) - 67 _____ feet tall!

5. Bulls have huge antlers which may reach (3 x 3) - 3 ____ feet across.

6. A moose can live as long as (35 -5) - 3_____ years.

7. Newborn moose weigh up to (5 x 14) ÷ 2 _____ pounds.

8. A cow, or female moose has (6 + 4) - 8 _____ calves each spring.

9. Moose need (80 ÷ 2) ____ pounds of food to eat, each day.

Did you know?

- In the former Soviet Union, moose have been domesticated as farm animals for milk.
- Sweden has the largest numbers of moose. One hundred thousand moose are killed each year for their furs and for meat.
- Moose prefer to live in forested areas that are near water.

Picture this!

- When mosquitoes and black flies annoy them, moose will coat themselves in mud or submerge themselves almost completely in a lake or pond.
- A moose can hold its head underwater for quite a long time while it gathers submerged vegetation with its fleshy lips.

Chapter Fourteen
Bears

Children (and adults) are awed, fascinated, and terrified by wild bears, but they love cuddly "Teddy" bears. All this emotion means that bears are wonderful, motivating subjects for reading, writing, and math!

There are three species of wild bear in North America: black bear, grizzly bear, and polar bear.

The black bear is the most widely distributed across the U. S. It is still found in good numbers where there are large tracts of forest or swamps. Its colors can vary from tan to brown to black, but in most areas it is black with a tan nose and a white spot on its chest. An adult black bear can be quite large—have your students research its size—but a newborn black bear is tiny, smaller than a squirrel.

> *"When a pine needle falls in the forest, the eagle sees it; the deer hears it, and the bear smells it."*
> *–an old First Nations saying*

The grizzly bear is listed as a "threatened" species on the U. S. Endangered Species list. A threatened species is one likely to become endangered unless strong measures are taken to increase its population. An endangered species is one which will become extinct unless strong measures are taken to insure its survival. The grizzly bear's name comes from the "grizzled" look caused by brown outer hairs with a lighter tip. Formerly found throughout the Western mountains of North America, it is now found in only 2 percent of its former range. Elsewhere, people have killed off the grizzly bear.

The polar bear is yellowish-white, and is found only in the North Polar regions, an area of Arctic tundra and pack ice. It is more carnivorous than the other two species, and eats mostly seals.

The grizzly bear is now found in only two percent of its former range.

In some places–especially our national parks–bears have become accustomed to park visitors who disregard rules against feeding the bears. When people disregard the rules, bears become a real nuisance and can become dangerous.

In some parks, bears have become very clever burglars, using their sharp, curved claws to pull out a car's window and get food inside. Bears recognize the shape of a cooler and know there is food inside. They will break into the car to get at the cooler.

Bears have a very good sense of smell. A package of chewing gum, or even a gum wrapper, gives off enough scent to interest a bear sniffing around the edge of the window or trunk, and then the bear will break in.

In national parks where bears are a problem of this magnitude, rangers usually warn visitors to be very careful, and educate visitors about what to do with their food. It is very important to protect your food. You will protect your own car from damage, as well as protect the bear from becoming a nuisance bear, a burglar bear, a fed bear, and a dead bear!

Activity 1: Compare Three Bears

Objective: Children will learn how to compare species systematically.

Concept: There is diversity within the bear family. Its members differ in their adaptations to their respective habitats.

For example:
–Polar bears are the best swimmers
–Black bears are best at climbing trees

You Will Need:
- *Writing implements*
- *Large pictures of all three North American bear species*

What to Do: Show children pictures of the three kinds of bears. Have children make a grid with rows and columns, decide how to compare the species, and research the species in books and on the Web (*Or, use the Student Handout Page provided on page 81.*)

The Student Handout Page lists some traits to research, but children can come up with their own ideas of traits to compare, too.

Did you know?
A black bear's claws are curved, an adaptation for climbing trees. A grizzly bear's claws are longer and straighter, better adapted for digging.

Activity 2: Trace the Range of Polar Bears Around the Top of the World

Objective: Children will learn that the range of animals transcends political boundaries, requiring national and international cooperation in their conservation. Children will learn that a globe is a three-dimensional map.

Concept: Like many animals of the Far North, the polar bear has a circumpolar range. It lives in northern Alaska, Canada, Greenland, and Russia. These countries cooperate in its conservation, usually permitting hunting only by indigenous (native) peoples.

You Will Need:
- *A globe*

What to Do: Let children find a two-dimensional map of the polar bear's range, then show their classmates on a turning globe where the polar bear lives.

Name

Compare Three Bears Worksheet

Look at pictures of the three kinds of bears. Research the species in books and on the Web.

•Use the information you find about about the different species to fill in the spaces below. Add add any interesting facts you find about bears on the back of the page.

	Black Bear	Grizzly Bear	Polar Bear
Color			
Length (adult)			
Weight (adult)			
Size of newborn young			
Range			
Food			
Climbs trees			
Size of ears			
Shape of claws			
Where cubs are born			
Safety tips if you live in this kind of bear country			

Activity 3: Make a Pie Chart

Objective: Children will learn what a bear eats, and how to symbolize a bear's diet by making a pie chart.

Concept: Children will research the food habits of North America's most numerous and widely distributed bear, the black bear. Pie charts can be fun for children to make with real objects. Children can readily understand that the biggest piece of the pie symbolizes what the bear eats the most of. Percentage information about a bear's diet can be hard to find; if children cannot find it in traditional sources, here is an educated estimate:

75%	**nuts:**	*acorns, beech, and hickory nuts*
	seeds:	*pine seeds*
	fruit:	*apples, cherries, blueberries, chokeberries, serviceberries, hawthorn berries, and grapes*
	grasses:	*and other plants*
10%	**meat:**	*squirrels, chipmunks, woodchucks, rabbits, and fawns*
7%	**insects:**	*ants and beetles*
5%	**fish:**	*suckers*
3%	**misc:**	*including honey*

You Will Need: • *A collection of objects symbolizing what a black bear eats, described below:*

What to Do: Real or simulated objects in a bear's diet can be attached in various ways– (glue, a loop of string, staple, etc.) to a circle of poster paper:
acorns
beechnuts
berries (sometimes plastic decorative ones can be purchased inexpensively)
pine seeds
roots (like a carrot)
the smallest possible container of honey
a small, realistically simulated fish (without a hook) which can be bought inexpensively in an angler's bait shop
chipmunks, etc. which the children can make out of construction paper or felt

Activity 4: Make Bear Tracks to Bear Food

Objective: Children will learn the approximate size and shape of black bear tracks, and also the pattern the tracks make as the bear walks.

Concept: Bear tracks are big and impressive, and the pattern of tracks is distinctive.

You Will Need:
- *A roll of paper at least 14 inches wide (because that is the width of a bear's set of tracks)*
- *A guide to animal tracks (optional)*
- *The bear tracks pattern on page 84*

What to Do: Have children cut the shape of bear tracks, front and rear, from black construction paper, using the pattern on page 84. They can then glue them onto a long, 14-inch-wide strip of paper which goes across the room and ends in front of the pie graph of a bear's food. Older children can apply their measuring skills, making their tracks the same dimensions as in the guide to animal tracks, and placing the footprints the same distance apart as shown in the tracking guide.

Activity 5: "A Fed Bear Is a Dead Bear"

Objective: Children will learn what causes "problem bears." They will learn about human safety, respect for bears, and bear safety.

Concept: The usual situation causing bear problems is easy access to a food supply. This changes the bear from a wary, wild animal to a dangerous beggar, dependent on humans and unafraid of them. People should never feed bears or leave food where bears can get it. Park rangers have a saying, "a fed bear is a dead bear." Any bear that learns it can get easy food by raiding campsites or dumpsters will end up causing enough trouble so that the bear will sooner or later have to be shot. People like bears so they feed them, either on purpose or out of ignorance, and ironically, it's very bad for the bear. A rare, quick glimpse of a wild bear at a distance is a thrilling sight. A beggar bear is a sad sight.

You Will Need:
- *Bear safety information and safety rules, available from any park where bears are found*

What to Do: Teach older children respect for wild bears. Remember that younger children will find the very thought of wild bears terrifying. It may help and reassure them to learn a few "bear facts," and some simple safety rules.

Children will enjoy role-playing different scenarios. One child could be the park naturalist or forest ranger, explaining what might happen if you feed a bear. Other children could play the role of tourists feeding junk food to an irascible bear (played by another child). Other children could act out the role of people respectfully watching bears from a distance. The "bear" would then behave naturally, for instance, eating blue berries. Children could talk very quietly, and then leave the area as the bear continues to eat blueberries.

Name

Making Tracks

Bear tracks are big and impressive, and the pattern of tracks is distinctive.

- **Cut the shape of bear tracks, front and rear, from black construction paper.**
- **Then, follow the directions given in Activity 4, on page 83.**

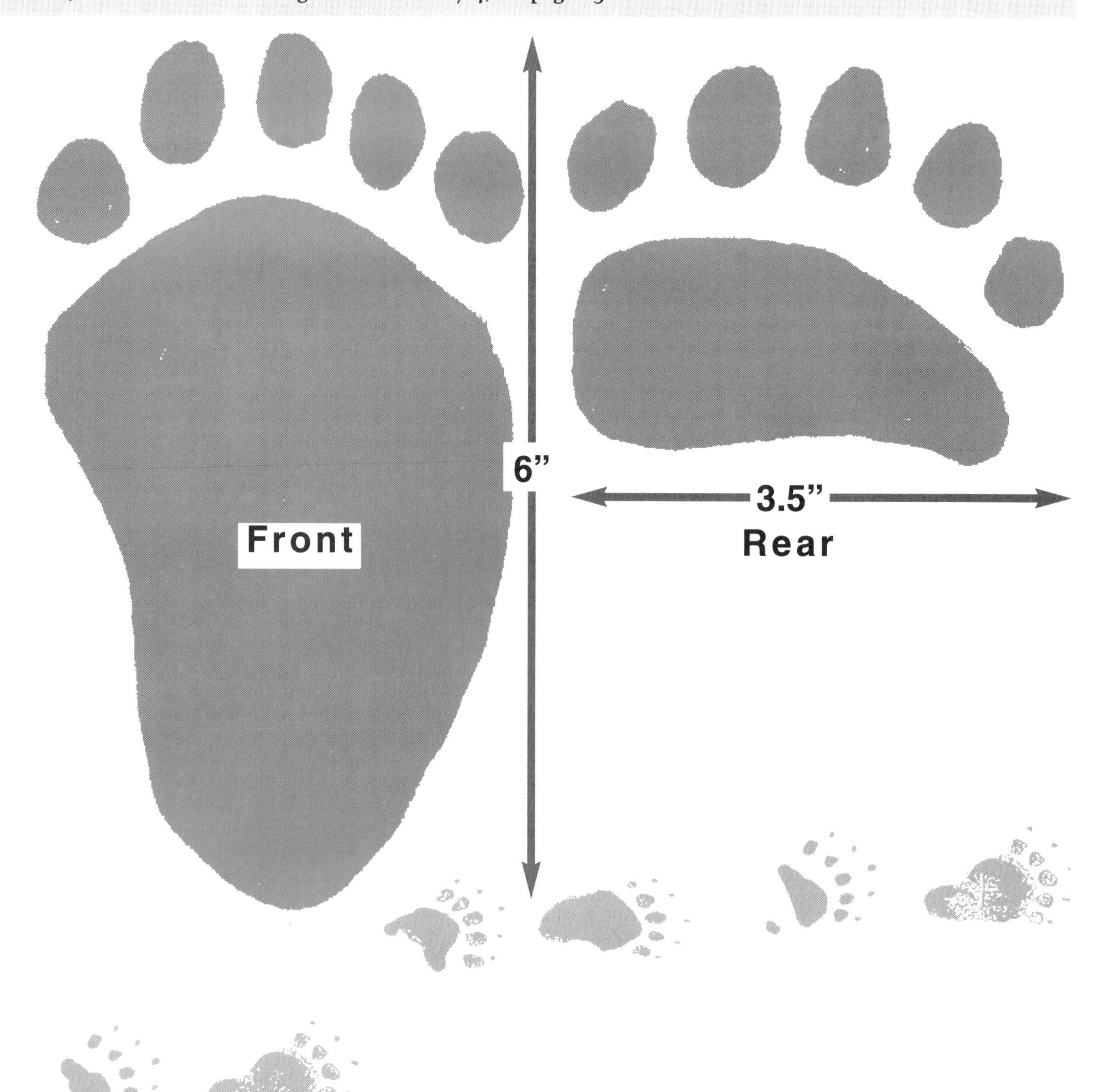

CHAPTER FIFTEEN

Now, What Is a Forest?

Remember the simple definition your students came up with when you first asked them, "What is a forest?"

Since then, they have learned a lot about the forest: from you, from the story *Shelterwood*, from other books, from the World Wide Web, from experiences and activities. Now it's time to ask them to expand on their earlier definition of a forest.

> *We are tired, never shaken, over-civilized people who are finding Nature is a necessity, that wilderness is freedom, that parks are fountains of life.*
>
> —*John Muir*

Activity 1: Make a Diagram of What You've Learned About the Forest

Objective: Children will learn to organize forest concepts, and to reinforce their own learning.

Concept: Organizing facts into categories is an important learning skill. Random, unrelated facts are hard to remember. But if they are related and organized into categories, an entire category with all its related ideas can be remembered as a whole.

You Will Need: • *A large writing board or easel, visible to all*

What to Do: Ask the children what they remember or like about the forest.
Write down their words, spaced evenly all over the writing board or easel.
Then say, "It's easy to forget so many different and wonderful things. Let's see if we can organize these ideas. That will make it easier to remember them."

Children can first learn to categorize with concepts like:

TREES—age, size, seedling, growth ring, species, etc.;
WOOD PRODUCTS—plywood, boards, linoleum, violins, etc.;
WILDLIFE—species, signs, homes, etc.;
SOIL—sandy, clay, wet, moist, color, etc.; and so on.

- Check chapter headings for more organizing concepts.
- Children can make a diagram or "concept map" of how these concepts are related.
- As children get older, the concepts they organize can get more abstract and complex.

It's a good feeling to get closure on a project and move on, but learning about the forest can also be a wonderful, lifelong adventure with no end.

A Glossary of Terms *as they are used in Forestry*

Advance regeneration—the new young trees that are already growing underneath the mature trees in advance of a harvest operation. These tiny trees will be the future forest after the mature trees are cut.

oak seedling

Annual ring—the layer of new wood which a tree grows around its trunk and branches each year in a temperate climate.

Biodiversity—the variety and variability of living things. It consists of many species, many genes, many habitats, many ecosystems.

Bumper trees—trees alongside a skidder path which protect the trunks of more valuable trees from scarring and damage as logs are hauled out of the forest.

Canopy—the upper layer of a forest, consisting mostly of foliage that receives direct sunlight.

Carbon—a chemical element which is an important constituent of all living things, including wood.

Carbon dioxide—a molecule consisting of carbon and oxygen.

Clearcut—*(noun)* an area in which all the trees have been cut down; *(verb)* to cut down all the trees in an area.

Composite—(as in wood composite) a wood-like substance made with resin and wood products like wood chips, fibers, or sawdust.

Coniferous—(as in coniferous tree) bearing cones.

Deciduous—(as in deciduous tree) shedding leaves on an annual basis.

Even-aged stand—a stand of trees which are all the same age. See Stand.

Evergreen—having foliage that is green throughout the year.

Forester—a trained professional in forest management: e.g., decision-making about harvest types and post-harvest growing of a new forest.

Growth ring—the layer of new wood which a tree grows around its trunk and branches each year in a temperate climate.

Hardwood—a broad-leaved tree such as an oak or maple; its wood is usually harder and denser than that of a softwood.

Harvester—a machine, operated by a person in its cab, which can cut and process trees mechanically, usually faster and more safely than they can be processed by a person with a chain saw and skidder.

Herbaceous—(as in herbaceous plant) soft-stemmed and not woody.

Intolerant—(as said of a tree) not able to survive in shade; intolerant of shade.

Linoleum—a kind of flooring made with linseed oil, pine resin, and wood flour.

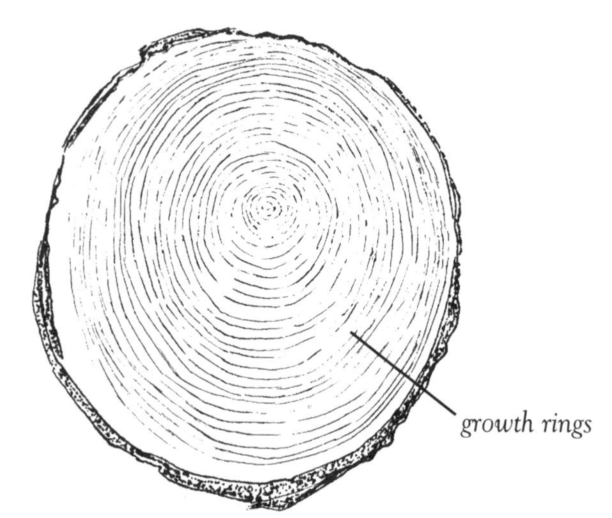

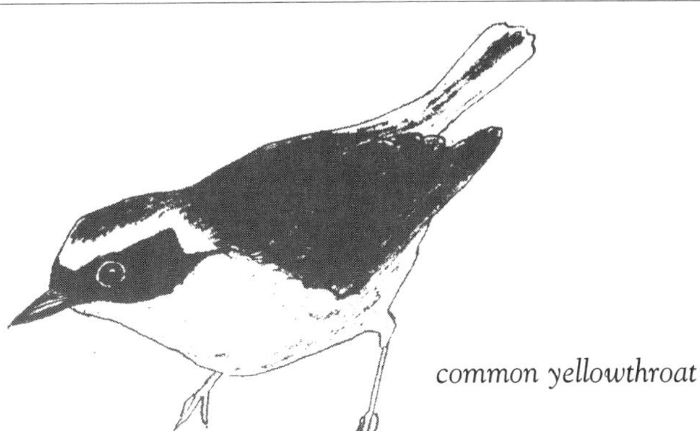

common yellowthroat

Old growth forest—a forest of complex structure in which the trees are very old, have never been cut, and in which there are many standing and fallen dead trees in addition to live trees.

Oxygen—a gaseous element necessary for respiration.

Photosynthesis—the chemical process of making food from carbon dioxide and water using energy from the sun.

Plywood—a wide board made by gluing together thin layers of wood, with the grain of each layer at right angles to the grain of the next layer. This makes plywood very strong.

Range—the area over which a plant or animal occurs.

Regeneration—the regrowth of the forest after cutting.

Rotation—the time allowed between successive cuttings of the same forest area (often 40–60 years).

Sapling—a small tree, its trunk less than 4 inches across.

Seedling—a tree newly grown from a seed, usually under 2 feet tall.

Seed tree—a healthy, valuable tree, selected to provide seeds to create a new generation of trees.

Selection Cutting—removal of selected mature trees, leaving high-quality, younger trees to grow.

Shade tolerance—(as said of a tree) the ability of a tree to survive in the shade of other trees.

Shelterbelt—a row or a line of trees that protect a house from wind.

Shelterwood—a harvest operation, often done in stages over many years, in which some trees are not cut down, but left standing to shelter new, young trees from extremes of heat, full sun, desiccation, the beating of heavy rain, and possible erosion. In addition to seedlings, other inhabitants of the forest floor, like wildflowers and salamanders, could also benefit.

Silviculture—the science of growing trees, by planting or by encouraging natural growth of young trees after cutting mature trees.

Skidder—a rubber-tired vehicle, something like a tractor, used to drag logs out of the forest.

Shrub—a woody plant with multiple stems, usually shorter than a tree.

Softwood—a cone-bearing tree with needle-like leaves; usually its wood is softer than that of a hardwood.

Stand—(noun) a group of trees all growing in a defined area, mostly of the same size and species. See Even-aged Stand, Uneven-aged Stand.

Succession—a predictable sequence of changes in species composition as a forest matures.

Sustained Yield—the amount of timber, and other forest products and services, that can be produced in perpetuity from a forest.

Tolerant—(as said of a tree) having the ability to survive in the shade; tolerant of shade.

Top—(verb) to cut the top off a tree (because it is too narrow to be of commercial value).

Twitch—(verb) to pull a bundle of logs from where they were felled to a yard; (noun) a bundle of logs.

Understory—the lower parts of trees, or the lower trees, that grow in the shade of the canopy.

Uneven-aged Stand—a stand of trees of different ages and sizes. See Stand.

Veneer—a thin, attractive layer of wood, often overlaid on wood composite.

Wilderness—a wilderness area is a large area where the human influence (roads, motorized vehicles, development) is absent.

Windbreak—a row or a line of trees that protect a house from wind.

Windthrow—a tree or group of trees that has been uprooted by the wind.

References *Background Reading for Adults*

Benyus, Janine M. *The Field Guide to Habitats of the Eastern United States*. Simon and Schuster, 1989.

Benyus, Janine M. *The Field Guide to Habitats of the Western United States*. Simon and Schuster, 1989. These two useful guides walk you through the major forest types, showing you the dominant tree species and the typical wildlife species of each, and even which "layer" of the forest they inhabit.

Berger, John J. *Understanding Forests*. San Francisco, CA: Sierra Club Books, 1998.

Davis, Mary Byrd, ed. *Eastern Old Growth Forests: Prospects for Rediscovery and Recovery*. Washington, DC: Island Press, 1996.

Devall, Bill. *Clearcut: The Tragedy of Industrial Forestry*. San Francisco, CA: Sierra Club/Earth Island Press, 1994.

Hammond, Herb. *Seeing the Forest Among the Trees: The Case for Wholistic Forest Use*. Vancouver, B.C.: Polestar Press, 1991.

Luoma, Jon R. *The Hidden Forest: The Biography of an Ecosystem*. Henry Holt, 1999

Maser, Chris. *Sustainable Forestry: Philosophy, Science, and Economics*. DelRay Beach, FL: Saint Lucie Press, 1994.

Pilarski, Michael, ed. *Restoration Forestry: An International Guide to Sustainable Forestry Practices*. Durango, CO: Kivaki Press, 1994.

Robinson, Gordon. *The Forest and the Trees: A Guide to Excellent Forestry*. Washington, DC: Island Press, 1988.

Sutton, Myron. *Eastern Forests*. Audubon Society Nature Guide. Alfred A. Knopf, Inc., 1997.

Whitney, Stephen. *Western Forests*. Audubon Society Nature Guide. Alfred A. Knopf, Inc., 1997.

Curriculum Guides

Project Learning Tree. American Forest Foundation, Washington, DC. A Program of the American Forest Foundation, *Project Learning Tree* (PLT) is a supplementary environmental education program. Teachers attend a one-day workshop where they receive a book of excellent activities for forest studies at all grade levels. Many of the activities are tried out at the workshop. The PLT curriculum provides supplementary activities about forest issues in subject areas such as social studies, language arts, mathematics, science, and art.

Trees are Terrific! Ranger Rick's *Naturescope*. National Wildlife Federation, 1998. What role do trees play in the "forest community"? How do trees differ from other plants? Why are they essential to the survival of the planet? These questions and more are answered in this teacher's guide that also teaches about rain forest conservation, community forestry, and what happens after a forest fire. Trees make the oxygen we breathe, cut pollution, prevent erosion, and provide food and shelter for both animals and people. Through interactive indoor and outdoor activities that show how to plant and protect trees, *Trees Are Terrific!* teaches about numerous tree species, leaves and trunk rings, why we need forests, and finally, how to celebrate trees!

Red squirrel tracks

References *Books for Children*

Arnosky, Jim. *Crinkleroot's Guide to Knowing the Trees*. Simon and Schuster, 1992. Explains differences between hardwood and softwood trees.

Beane, Rona. *Backyard Explorer Kit*. Workman Publishing Co., 1989. This brings the joy of discovery to children, guiding them in collecting leaves and exploring trees. Suggests leaf and tree projects for the entire year.

Bowen, Betsy. *Tracks in the Wild*. Houghton-Mifflin, 1998.

Bunting, Eve. *Someday a Tree*. Clarion Books, 1996. A dying oak tree lives on in its acorns.

George, Lindsay Barrett. *In the Snow: Who's Been Here?* Greenwillow Books, NY, 1995. This is a beautifully illustrated picture book, with minimal text, about reading animal "signs."

Greeley, Valerie. *The Acorn's Story*. Macmillan Co., 1994. This book illustrates the fascinating transformation from acorn to full-sized oak.

Jasperjohn, William. *How the Forest Grew*. Mulberry Books, 1992. This excellent book shows forest succession happening before your eyes, yet is simply written and understandable for the picture book set or for young readers.

Lauber, Patricia. *Be a Friend to Trees*. HarperCollins Children's Books, 1994.

Lyon, George Ella. *A B Cedar: An Alphabet of Trees*. Demco Media, 1996.

Maestro, Betsy. *Why Do Leaves Change Color?* Demco Media, 1994.

Robbins, Ken. *Autumn Leaves*. Scholastic Trade, 1998

Romanova, Natalia. *Once There Was a Tree*. E. P. Dutton, 1992. An old stump attracts many living creatures.

Udry, Janice May. *A Tree's Nice*. Harper Trophy, 1987. A Caldecott Award winner and a children's classic, this presents all the wonderful things trees do.

Yolen, Jane. *Owl Moon*. Philomen Books, 1987. A grandfather and granddaughter go on an "owl prowl" after dark.

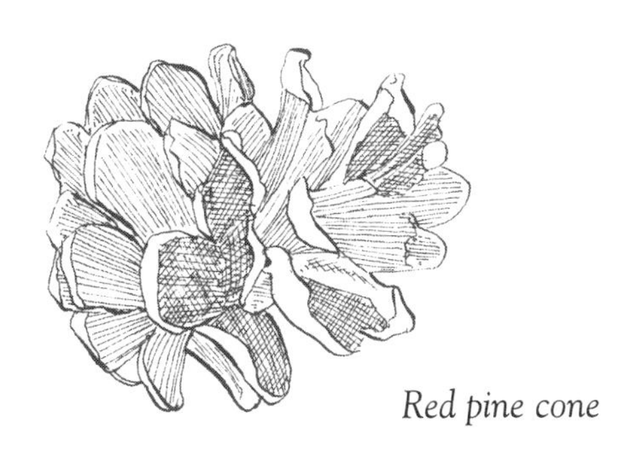

Red pine cone

References *Websites for Children*

Organizations with Useful and Appropriate Websites for Children:

American Forests
P.O. Box 2000
Washington, DC 20013
(202) 955-4500
http://www.amfor.org
Website features *Historic Trees* and *Big Trees.*

Forest Service Employees
for Environmental Ethics
P.O. Box 11615
Eugene, OR 97403
(541)484-2692
http://www.afseee.org
Website features *The Secret Forest* (for children) and also a *Forest Dictionary.*

The Wilderness Society
900 17th St. NW
Washington, DC 20006-2506
http://www.wilderness.org

Eastern chipmunk

Answer Key

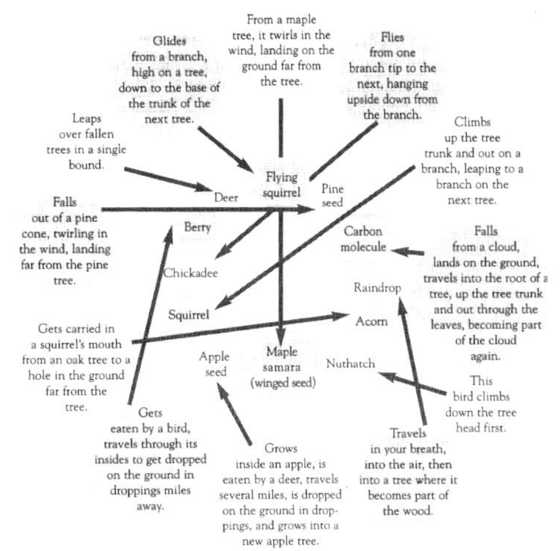

p. 10 Woodland Wanderers

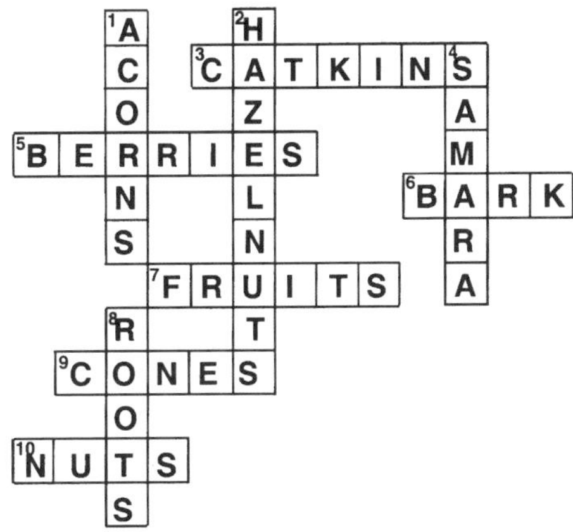

p. 11 Fruits of the Forest

p. 12 Forest Friends

1. One tree can provide a day's oxygen for up to four people.
2. Young bark is smooth. Old bark cracks and flakes.
3. Counting the rings on a section of trunk can tell us the age of a tree.
4. There are about 1000 species of trees in North America.
5. Trees are the largest living things in the world.
6. A young tree is called a sapling.
7. Bark grows from the inside and pushes the older bark outward.
8. Every inch around the girth of a tree corresponds to a year in a tree's growth.
9. More than 400,000 leaves can falls from a single large tree.
10. The tallest tree in the world is a California redwood and is over 360 feet tall.
11. The bristlecone pine is the oldest living tree at over 4,600 years old.

p. 17 Tree Trivia

Crack the Code

Riddle: What kind of pine has the sharpest needles?

Answer: A PORCUPINE

p. 18 Crack the Code

p. 30 Leaf It to Me

Crack the Code

Riddle: If a mouse loses its tail, where can it get a new one?

Answer: AT A RETAIL STORE

p. 34 Crack the Code

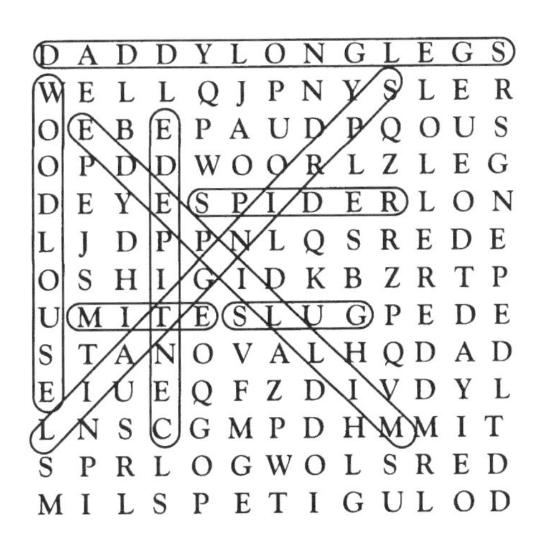

p. 46 Litter Bugs

Crack the Code

Riddle: What do you call three singing trees?

Answer: A TREE-O

p. 51 Crack the Code

1. **ANTLINGS**
2. **CUBS**
3. **NESTLINGS**
4. **CALVES**
5. **PUPS**
6. **FAWNS**
7. **EAGLET**
8. **TADPOLES**
9. **LARVA**
10. **OWLET**
11. **KITTEN**
12. **CHICKEN**
13. **BUNNIES**

p. 52 Babes of the Forest

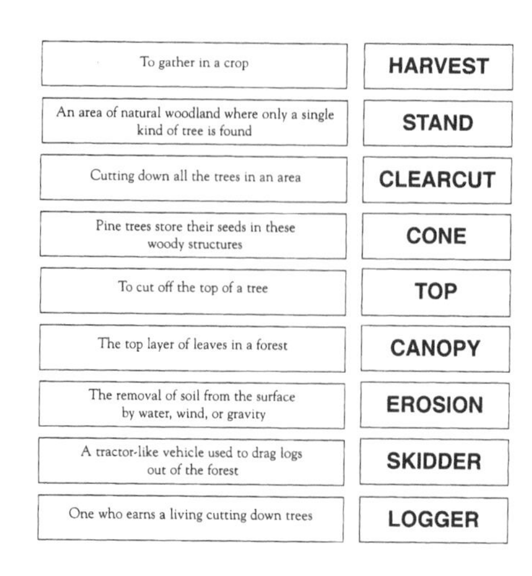

To gather in a crop	**HARVEST**
An area of natural woodland where only a single kind of tree is found	**STAND**
Cutting down all the trees in an area	**CLEARCUT**
Pine trees store their seeds in these woody structures	**CONE**
To cut off the top of a tree	**TOP**
The top layer of leaves in a forest	**CANOPY**
The removal of soil from the surface by water, wind, or gravity	**EROSION**
A tractor-like vehicle used to drag logs out of the forest	**SKIDDER**
One who earns a living cutting down trees	**LOGGER**

p.56 Woods Words

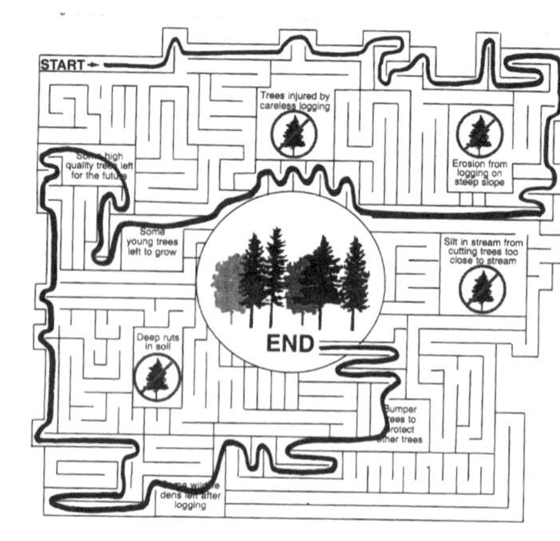

p. 58 AMazing Forests

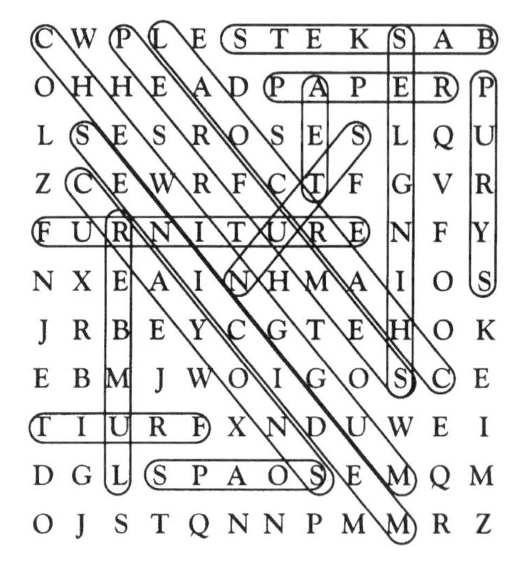

p.61 Products We Get from Trees

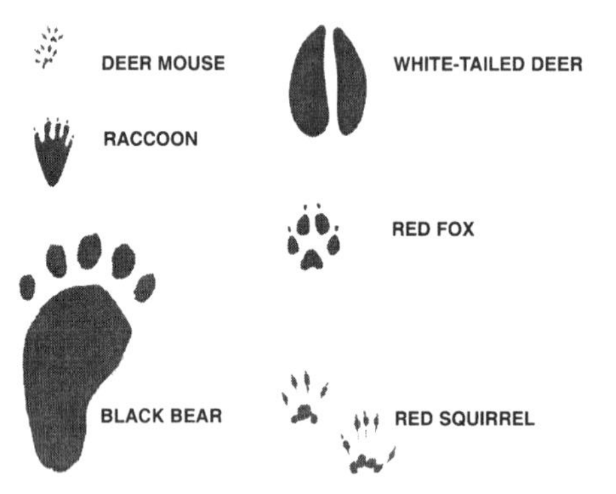

p. 69 Tricky Tracks

c m e i	M I C E
r e d n m a s l a a	S A L A M A N D E R
t s c e i n s	I N S E C T S
m e r a h t o w r	E A R T H W O R M
e l o v	V O L E
n k s a e	S N A K E

p. 70 Life in the Leaves

1. The moose is the largest member of the deer family, and may weigh over (500 x 3) - 100 **1400** pounds!
2. Moose can run up to (6 x 5) + 5 **35** miles an hour.
3. The tracks of the bull, or male moose, can be (4 x 3) - 5 **7** inches long.
4. At five months old, the moose calf may stand (6 x 12) - 67 **5** feet tall!
5. Bulls have huge antlers which may reach (3 x 3) - 3 **6** feet across.
6. A moose can live as long as (35 -5) - 3 **27** years.
7. Newborn moose weigh up to (5 x 14) ÷ 2 **35** pounds.
8. A cow, or female moose has (6 + 4) - 8 **2** calves each spring.
9. Moose need (80 ÷ 2) **40** pounds of food to eat, each day.

p.78 Moose Math

Judy Kellogg Markowsky is the director of Maine Audubon's Fields Pond Nature Center, has an Ed.D. in Science Education, is the coauthor of the *Everybody's Somebody's Lunch Teacher's Guide*, and has created award-winning environmental education programs, including many forest ecology tours and classes for every age group.

Rosemary Giebfried is a designer and illustrator with a special interest in science. She also designed and illustrated the *Everybody's Somebody's Lunch Teacher's Guide*.

Books for the Classroom, from Tilbury House

Tilbury House has developed an outstanding series of award-winning children's picture books—books that explore cultural diversity, nature, and the environment. They are wonderful read-together books for the entire family, but also used by teachers in schools across the country.

Working with educators who are specialists in their fields, we have developed imaginative teacher's guides with hundreds of ideas for cross-curriculum units. Many of our books are approvied for the California Instructional Materials Fund, the New York State Textbook List, and have been adopted in New Mexico. Our newest books and teacher's guides will be submitted to the California and New York lists as soon as they are in print.

Most of our authors offer workshops for students and teachers and are also available for booksignings in bookstores. Please call us for more information. We also publish a free newsletter about using our books in the classroom, and it's available upon request.

We welcome your comments and suggestions about our books, and we look forward to bringing you new books each year that offer special learning experiences, spark the imagination, and open new worlds for young readers.

About Our Other Books on Nature • Science • Ecology

Sea Soup: Phytoplankton

Mary M. Cerullo
Photography by Bill Curtsinger

Published with the Gulf of Maine Aquarium
Hardcover, $16.95; ISBN 0-88448-208-1
9 x 10, 40 pages, color photographs
Children/Science; Grades 3–7

A teaspoon of sea water can hold a million phytoplankton! These tiny floating plants come in thousands of amazing variations, including chains, spirals, boxes, and spiky balls. Their tiny individual size belies their importance. Invisible to the naked eye, phytoplankton are the source of our atmosphere, our climate, our ocean food chain, much of our oil supply, and more. When phytoplankton bloom, they can turn the sea into a thick phytoplankton soup that directly or indirectly feeds everything else in the ocean. They remove half of all the carbon dioxide that humans release into the atmosphere each year from burning oil and gas, at the same time creating more oxygen for us to breathe. Yet despite all these benefits, some phytoplankton can be deadly—causing outbreaks of "red tide" that poison those who eat shellfish.

Bill Curtsinger's extraordinary photomicroscopy serves up tantalizing images of this "sea soup" while Mary Cerullo's text answers intriguing questions about these tiny drifters that have shaped our world.

Mary Cerullo is a children's science writer and the author of *Dolphins: What They Can Teach Us*, *The Octopus: Phantom of the Sea*, *Reading the Environment: Children's Literature in the Science Classroom*, *Coral Reef: A City That Never Sleeps*, *Lobsters: Gangsters of the Sea*, and *Sharks: Challengers of the Deep*. She lives in South Portland, Maine, and is assistant director of the environmental organization Friends of Casco Bay.

Bill Curtsinger's photography has appeared in numerous books and magazines, including *National Geographic*, *Life*, *Time*, *Newsweek*, *Outside*, *Natural History*, and *Smithsonian*. He has photographed three books, *Wake of the Whale*, *The Pine Barrens*, and *Monk Seal Hideaway*. His photomicroscopy for this book was done at Maine's Bigelow Laboratory for Ocean Science, which holds the world's largest collection of phytoplankton. The images were done on an Axiophot 2 research microscope donated by Carl Zeiss, Inc. Bill lives in Yarmouth, Maine.

Sea Soup Teacher's Guide

Discovering the Watery World of Phytoplankton and Zooplankton

Betsy T. Stevens
Illustrated by Rosemary Giebfried

Published with the Gulf of Maine Aquarium
Paperback, $9.95; ISBN 0-88448-209-X
8 1/2 x 11, 96 pages, illustrations
Children/Science; Grades 3–7

The interesting and fun activities in this teacher's guide meet the challenge of relating tiny, microscopic organisms to the lives of children. Discover and explore answers to some strange questions. What *is* the recipe for Sea Soup? Are those tiny critters plants or animals, or maybe something else? Why do they look more like creatures from outer space than the organisms we know on land? What do giant clams, corals, whales, penguins, and humans have in common? How does the Sea Soup grow? What if it stops growing?

The inquiry-based activities range from designing and making a phytoplankter and collecting plankton to designing an experiment for exploring what factors influence the growth of phytoplankton and zooplankton. The emphasis is on science, but where appropriate math, geography, language arts, and art are included. Each unit includes background information, objectives, a statement of how it addresses the National Science Education Standards, materials, procedures, references, and suggested websites.

Betsy T. Stevens grew up mucking about in mudflats and salt marshes on the Connecticut shore. Following graduate school at Cornell, she taught biology at Skidmore College in upstate New York for twenty-five years, and then became the director of Sandy Point Discovery Center on the Great Bay Estuary in New Hampshire. Recently retired, Betsy lives in Kennebunk, Maine, where she writes (primarily school curriculum materials), sails, explores, and volunteers.

Sea Soup: Zooplankton will be published in 2000!

STONE WALL SECRETS
Kristine and Robert Thorson
Illustrated by Gustav Moore
Hardcover, $16.95; ISBN 0-88448-195-6
9 x 10, 40 pages, color illustrations
Children/Science; Grades 3–7

• "Notable Books for Children, 1998" —*Smithsonian*
• "This heartfelt picture book features a well-crafted story, detailed descriptions, and distinctive paintings. After reading this outstanding publication, readers will never look at old stones the same way again." —Debra Briatico, *Children's Literature*
• *Stone Wall Secrets* is a book chock-full of real geology, in plain English that will make sense to children and laymen alike...this book will appeal to children and to anyone who wants children to love learning science. —*Appraisal*

What can the rocks in old stone walls tell us about how the earth's crust was shaped, melted by volcanoes, carved by glaciers, and worn by weather? And what can they tell us about earlier people on the land and the first settlers? As Adam and his grandfather work together to repair the family farm's old stone walls, Adam learns how fascinating geology can be, and how the everyday landscape provides intriguing clues to the past. *Stone Wall Secrets* also shows positive family dynamics between different generations and different races in an adoptive family. Gus Moore's richly detailed paintings are the perfect complement to a story full of imagery and wonder.

Robert Thorson is a professor of geology and geophysics at the University of Connecticut in Storrs, where he holds a joint appointment in anthropology. His expeditions have taken him from Alaska to Chile, and have included excavating woolly mammoths and "cave men," and mapping glaciated wilderness areas. Kristine Thorson's special interests are language and intercultural understanding. Illustrator Gustav Moore graduated from the Rochester Institute of Technology with a degree in illustration and now lives in Portland, Maine.

STONE WALL SECRETS TEACHER'S GUIDE
EXPLORING GEOLOGY IN THE CLASSROOM
Ruth Deike
Paperback, $9.95; ISBN 0-88448-196-4
8 1/2 x 11, 90 pages, illustrations
Education/Science; Grades 3–7

Ruth Deike, a geologist with the U. S. Geologic Survey for more than thirty years and the founder of The Rock Detective, a non-profit educational organization, brings boundless energy to teaching school children about earth science. Her Teacher's Guide incorporates the imagery and wonder of *Stone Wall Secrets* with hands-on classroom activities that illustrate basic earth science concepts, opening doors into ideas and concepts more beautiful and wild than Star Wars, Star Trek, and Superman combined! Ruth has learned that what young Rock Detectives discover for themselves, they remember. Working with the National Science Education Standards, she has created a variety of exciting activities, from exploring the earth's building blocks to studying volcanoes to posing intriguing questions such as, Does the Earth itch? Or, Will there be another ice age?

EVERYBODY'S SOMEBODY'S LUNCH
Cherie Mason
Illustrated by Gustav Moore
Hardcover, $16.95; ISBN 0-88448-198-0
9 x 10, 40 pages, color illustrations
Children/Nature; Grades 3–6

• "This is one of the best books I have seen on predator and prey."
—*Portals, Idaho Reading Journal*
• "Yes! Here's a story that takes us beyond wolves and bears and presents the important and influential role of many predators, including humans." —Tom Skeele, The Predator Project
• "Cherie Mason lets us feel the touch of a fox's nose in her wonderful first book, *Wild Fox*. Here she tells the difficult but important tale of predation in a format children will understand and accept."
—*Paws in Print*

Many children—indeed, many adults—believe that there are "good" animals and "bad" animals. The Big Bad Wolf myth lives on. This new story puts predators in an entirely new light as a sensitive young girl, shocked and confused by the death of her cat, learns the roles that predator and prey play in the balance of nature. Gently and gradually, she comes to understand why some animals kill and eat other animals in order to live. It is one of nature's most exciting and important lessons. Children and all who read to them will come away with a new respect for all wildlife. In keeping with our commitment to diversity education, this story also shows an extended family rich in racial and cultural diversity.

Cherie Mason has been active in environmental and wildlife protection causes for more than thirty years and is an environmental radio journalist. She is the author of the award-winning children's book *Wild Fox: A True Story* and makes her home on Deer Isle in Maine. Gustav Moore graduated from the Rochester Institute of Technology with a degree in illustration and now lives in Portland, Maine. He is also the illustrator for *Stone Wall Secrets*.

EVERYBODY'S SOMEBODY'S LUNCH TEACHER'S GUIDE
THE ROLE OF PREDATOR AND PREY IN NATURE
Cherie Mason and Judy Kellogg Markowsky
Illustrated by Rosemary Giebfried
Paperback, $9.95; ISBN 0-88448-199-9
8 1/2 x 11, 70 pages, illustrations
Education/Nature; Grades 3–6

The important roles that predator and prey play in the balance of nature are gently explained to children in Everybody's Somebody's Lunch. This Teacher's Guide provides educators with information, activities, and play that can easily be incorporated into wildlife and nature study programs. Included are the history of the persecution of predators due to human ignorance and fear; profiles of predatory mammals, invertebrates, reptiles, amphibians, birds, and marine life; humans as predators; and hopeful evidence of change in today's attitudes. These critical environmental lessons are structured so that they are interesting, instructive, and fun.

Judy Markowsky is the director of Maine Audubon's Fields Pond Nature Center, has an Ed.D. in Science Education, and has created award-winning environmental education programs.

Project Puffin

How We Brought Puffins Back to Egg Rock

Stephen W. Kress, as told to Pete Salmansohn

Hardcover, $16.95; ISBN 0-88448-170-0
Paperback, $7.95; ISBN 0-88448-171-9
10 x 7 1/4, 40 pages, color photos
Children/Nature, Grades 3–6

- A National Audubon Society Book
- "Notable Books for Children, 1997" —*Smithsonian*
- "Outstanding Science Trade Books for Children for 1998" —Children's Book Council
- "Way to go, Project Puffin! What a great story about a person doing something to help our fellow creatures. Welcome back, puffins!" —Martin Kratt, *Kratts' Creatures*

With their large, colorful beaks, their upright posture, and their big, dark eyes, it's easy to see why puffins are popular all over the world. But for the past hundred years, puffins along the coast of Maine have been threatened with local extinction. Biologist Stephen Kress decided to try to bring puffins back to Maine with an experiment that had never been attempted before.

Stunning color photographs on every page capture each step of this wildlife success story. As you learn about The Puffin Project, you'll also learn all about puffins—how they are so wonderfully adapted to their ocean environment, how they catch fish, socialize, nest in burrows, and raise their young. *Project Puffin* is a fascinating wildlife study as well as the true story of a young scientist with a hopeful dream.

Giving Back to the Earth

A Teacher's Guide to Project Puffin and Other Seabird Studies

Pete Salmansohn and Stephen W. Kress
Illustrated by Lucy Gagliardo

Paperback, $9.95; ISBN 0-88448-172-7
8 1/2 x 11, 80 pages, illustrated
Education/Nature; Grades 3–6

- A National Audubon Society Book
- "Educationally sound, it offers a wide variety of experiences to enhance and enrich the student's understanding of puffins, seabirds, and the oceans.... This is a marvelous and fun resource." —*Appraisal*

Here are more than 40 creative, hands-on activities: art projects, role-playing, wildlife observations, science demonstrations, running games, and more. The guide is organized into seven major themes, including seabird adaptations, the marine ecosystem, human impact on the environment, people making a difference for wildlife, and more. Includes annotated bibliographies and Internet resources.

Stephen Kress, Ph.D., is the National Audubon Society's seabird expert and has directed The Puffin Project since 1973. Pete Salmansohn has worked with Steve since 1980, has taught thousands of children and led hundreds of public boat cruises to see puffins on Eastern Egg Rock, Maine, and was chosen Maine Environmental Educator of the Year in 1998.

Tilbury House, Publishers

2 Mechanic Street • Gardiner, ME 04345

800-582-1899 • 207-582-1899 • Fax 207-582-8227 • E-mail tilbury@tilburyhouse.com